What if?
Mind-Boggling Science Questions for Kids

Robert Ehrlich

Illustrated by Ed Morrow

John Wiley & Sons, Inc.

New York • Chichester • Weinheim • Brisbane • Singapore • Toronto

This book is printed on acid-free paper. ∞

Copyright © 1998 by Robert Ehrlich.
All rights reserved.

Illustrations © 1998 by Ed Morrow. All rights reserved.

Design and production by Navta Associates, Inc.

Published by John Wiley & Sons, Inc.
Published simultaneously in Canada.

Library of Congress Cataloging-in-Publication Data:

Ehrlich, Robert
    What if? : mind-boggling science questions for kids / Robert Ehrlich
       p. cm.
    Includes index.
    Summary: Questions and answers explore the earth, weather and climate, forces and energy, plants and animals, and other scientific subjects.
       ISBN 0–471–17608-7 (pbk. : alk. paper)
    1. Science—Miscellanea—Juvenile literature. [1. Science—Miscellanea. 2. Questions and answers.]
    I. Title.
    Q173.E46 1998
    500—dc21
97–35863

Printed in the United States of America.

10 9 8 7 6 5 4 3 2 1

Acknowledgments

The following young people have read through an early version of this book and have made very helpful suggestions: Kristen English (age 12), Jilene Jackson (age 12), Jennifer Kallenborn (age 12), and Sarah Schanze (age 11). I am also especially grateful to my niece Richelle Adler, an elementary school teacher, for her kind and helpful suggestions on the book.

Contents

Introduction

Albert did not do very well in school. In fact, one of his teachers told him that he would never amount to anything. Albert used to daydream about the world around him and used to ask himself strange questions. He wondered, for example, what would happen if he shone a beam from a flashlight and then chased after the light. Would he begin to catch up with it if he could travel very fast? If he could travel as fast as the light itself, what would he see? Would the light appear to stop moving? Albert remembered this imaginary experiment when he grew up. It became the basis of his most famous theory—the theory of relativity. Albert Einstein (1879–1955) is now considered to have been one of the greatest scientists of all time. Among other discoveries, he invented the theory of relativity.

Albert Einstein's imaginary experiments and "what if" questions were an important way he learned about the universe. Many other scientists have also asked themselves "what if" questions to learn how the world works. Did you ever ask yourself: what if the Moon fell down? The scientist Isaac Newton asked himself that question 300 years ago, and it led him to discover his theory of gravity.

Science is not just a bunch of facts that you might find in some book. It's a way of learning about our world by asking questions, doing experiments, and seeing what happens. Scientists are the explorers of our world today. But even if you don't want to become a scientist, asking "what if" questions can be a lot of fun. Did you ever imagine what things would be like if you could see in the dark or if you could fly? Or if we tried to communicate with aliens? Or bring the dinosaurs back? These are just some of the amazing questions in this book. Each question is answered with a simple statement, followed by a longer explanation of the science behind the answer. Boxes within the answer provide fun facts, simple experiments you can do, and more interesting questions you might ask. By the end of

the book, you'll probably be thinking of your own "what if?" questions—questions that might lead you to future discoveries in science and maybe even change the way people think about the world.

Big, Blue, Spinning Ball

The Earth

S ay you throw an apple up in the air. What makes the apple fall—bonk—right back down? Gravity. **Gravity** is the force that pulls matter to other matter. The heavier and larger something is, the more gravity it has. The Earth is so large, it pulls everything else on it down—including you, the apple you throw up in the air, and the air itself. The layer of air around the Earth that we call the atmosphere is held down by gravity. Gravity also pulls the Earth and Sun together. But don't worry, the Earth is not going to fall into the Sun! The Earth goes around the Sun in a circular path we call its **orbit,** which takes a year to complete. The pull of gravity is what keeps the Earth in its orbit around the Sun. As the Earth orbits the Sun, it also spins or **rotates** on its **axis** (an imaginary line through its center) once each day.

What if you dug a hole through the Earth

If you didn't get crushed, you'd get really hot! The center of the Earth is constantly being squashed by the weight of all the outer layers on top of it that push inward from all directions. This squashing makes the Earth's center very hot—up to 6,000 degrees Fahrenheit (about 3,333°C). Ouch!

In fact, the middle of the Earth is so hot that the iron in the Earth's outer **core** (center) is liquid. Imagine trying to dig a hole through boiling water, and you can begin to see what digging a hole through the Earth would be like.

The deepest holes anyone has been able to make on our planet are only 9 miles (14 km) deep. The deepest holes don't even make it through the Earth's **crust**—its outer layer, which is about 20 miles (30 km) thick.

DOWN, DOO-BEE-DOO-DOWN-DOWN. Which direction is down for people living on the opposite side of the Earth from you? Since the Earth is a giant ball, are they in danger of falling off? No, because no matter where you are on this giant ball, the direction you call "down" points toward the center of the Earth. Gravity pulls everyone toward the center of the Earth. Look at the drawing. People on one side of the Earth are standing upside down, from someone's point of view. Then turn the drawing all the way around. Now the people who were rightside up are standing upside down. No one falls off the Earth. ◀

That isn't long at all when you consider that the distance from one side of the Earth to the other (the Earth's diameter) is 8,000 miles (12,800 km). If the Earth were a balloon the size of a basketball, the deepest holes on Earth wouldn't even go through the balloon skin.

Nobody's upside down.

What if you tried a special pipe?

Even if you used a special nonmelting pipe to shove through the Earth, you'd run into a bigger problem. The innermost core of the Earth is made of solid iron. You could never find an instrument strong enough to dig through it. And even if you did find an instrument strong enough, you'd have to fight the tremendous force of gravity that would surely squash it.

PRETTY BORING! *Geologists are the ones digging the deepest holes in the Earth. To collect samples of the rocks in the Earth's crust, they drill 9-mile-deep "boreholes" (to* **bore** *means to drill).*

What if
you dropped a rock down a hole through the Earth

Drop a rock down a deep well and nothing exciting happens. Eventually, you hear the rock go plunk. Now imagine dropping a rock through an imaginary hole through the Earth—that's a different story. Okay, you know from the previous answer that you couldn't really make a hole through the Earth because of its liquid center. But what if you could? The rock would fall faster and faster down the hole. It would be pulled by gravity all the way to the center of the Earth. As it passed the center, it would be moving VERY fast, even faster than the space shuttle!

Now, a funny thing would happen as the rock kept falling past the center. After it passed the center, it would get closer

What if you got a friend to catch the rock on the other side of the Earth?

I hope your friend is good at baseball. She'd have to act fast to grab the rock when it popped out of the hole. That's because the rock would just barely make it to the opening of the hole. The imaginary rubber band would then be stretched just as much as it was when you first let go of the rock.
If your friend didn't grab the rock in time, what do you think would happen? You guessed it! The rock would fall back all the way through the Earth again. Then you would have a chance to grab it when it popped back up on your side. The whole trip would take just about two hours. If neither of you grabbed the rock, it would just keep going back and forth through the hole like a yo-yo— the most amazing yo-yo *in* the Earth! Eventually, the rock would slow down and come to rest at the center of the Earth because of **friction.** Friction is the rubbing of one object against another object or against the air.

and closer to the surface on the other side of the Earth. The rock would really be moving up, not down! Soon the rock would begin to move more slowly. The Earth's gravity would pull it back toward the center, slowing it down.

Why? Think of gravity like the pull of a rubber band. Imagine that the rock is tied to one end of a long stretched rubber band. The other end of the rubber band is connected to the center of the Earth. When you first let the rock go, the stretched rubber band pulls the rock toward the center. As the rock passes the center and starts rising toward the surface on the other side of the Earth, the rubber band again pulls it back toward the center.

"Look for it in about two hours."

What if the Earth were square?

If the Earth were square, travel agencies could really sell trips to "the four corners of the Earth." The four corners would be like tall mountains. Can you picture it?

Okay. Let's imagine that the Earth looked like a square box, and that you were the size of an ant walking around the box. The direction "down" would always be toward the center of the box. If you were standing in the middle of one side of the box, the ground would look perfectly flat as far as you could see.

Now, suppose you started walking toward a corner of the box on your teeny-weeny little ant legs. As you got closer to the corner, you would feel like you were climbing a mountain that got steeper and steeper. Right at a corner you would be at the top of a mountain.

Is there some reason the Earth and all the other planets are round, not square? It's the big G—gravity. The force of gravity pulls everything on Earth toward the center. It also pulls the pieces of the Earth itself toward the center. Gravity pulls

I Got the Whole World in My Hands *How gravity can make a square planet round.* Take a square lump of modeling clay. Imagine that the palms of your hands are the force of gravity. Mold the clay in your hands by pushing every part of the clay toward the center just like the big G. If you do this for a while, the lump of clay will begin to become round. In fact, the piece of clay will get rounder and rounder the more time you spend pushing each part toward the center. Of course, if you pushed very gently on the square lump of clay it wouldn't change its shape much. In exactly the same way, if gravity were not so strong a force, it wouldn't be able to make a planet round if it started out as some other shape.

extremely hard. Even if the Earth started out square or in the shape of an octopus, the pull of gravity toward the center would quickly mold it into a **sphere** (a round ball)—but not a perfect sphere. Because of its rotation, the Earth bulges slightly around the middle.)

PLANETS X, Y, Z . . . *There are some really tiny planets (not counted among the usual nine) that have extremely weak gravity. Gravity on these planets, called **asteroids,** is so weak that the planets aren't round, but have some really weird shapes, like any rocks on Earth. Gravity on asteroids is just too weak to mold them into spheres.*

How about other planets that have different strength of gravity from Earth? The bigger something is, the more gravity it has. Giant planets such as Jupiter have very strong gravity, and tiny planets such as Pluto have very weak gravity. (Read all about what it would be like to live on the other eight planets in our solar system in "What if you visited Uranus or Neptune?" on page 138.)

9

What if you lived at the North Pole?

Your day would be six months long and your night would be six months long. During your six-month-long day you might want to play all the time, not sleep. But during the six-month-long night you might want to make like a bear and **hibernate** (become inactive). Zzzzzzz.

Why? The Earth makes one spin a day around its axis. This axis is slightly tilted in relation to the Earth's orbit around the Sun. The two points where the axis passes through the Earth are the North Pole and the South Pole. As the Earth spins, it also takes a one-year trip around the Sun. As it moves around the Sun, the North Pole is tilted toward the Sun for half the year. For the other six months it's tilted away from the Sun. So, at the North Pole, it's day for six months of the year and night for the other six months. (Check it out in the "Polar Expedition" experiment.)

If you stood at the North Pole and looked up at the sky, what would you see? It would look like the whole sky rotated around a point directly overhead once every 24 hours. To get the idea, look up at the ceiling and imagine that a Sun, a Moon, and stars are painted on it. If you spin yourself in a circle, it looks like everything on the ceiling is going in circles.

How come it's always cold at the North Pole and South Pole?

You'd expect it to be 60 degrees below zero during the six months of darkness. But why doesn't it get really hot during six long months of continuous daylight? Because at the Poles the Sun never gets very high in the sky. When the Sun is low, the Sun's rays always strike the ground at a small angle, causing very little heating.

Polar Expedition *How you get six months of daylight at the North Pole.* Put a small lamp in the middle of your kitchen table. The lamp represents the Sun. Hold a globe of the Earth at the same height as the lamp. Walk the globe around the Sun in a big circle while you spin the globe on its tilted axis. (The actual Earth spins on its axis 365 times for each trip around the Sun.) As you walk the globe around the circle, see when the light from the lamp reaches the North Pole and when the Pole is in shadow. You should find that the North Pole receives no light from the lamp for half the time around the circle—the six months of darkness. For the other half of the year—the six months of daylight—the North Pole gets continuous light from the lamp.

If you lived on the North Pole, during the six months of daylight you wouldn't see the Sun rise or set. The Sun would trace a circle in the sky every 24 hours. Toward the end of the six months of daylight, the Sun would be very close to the horizon. Then, it would dip below the horizon for the next six months, still traveling in circles each 24 hours.

What if the Earth had no magnetism?

You'd be likely to get lost on your next camping trip. Without the Earth's magnetism, compass needles wouldn't point north. We also wouldn't have North America, the South Seas, the East Coast, or the West Indies.

Have you ever played around with a pair of magnets? The two ends of a magnet are usually labeled N and S for north and south poles. If you bring two north ends near each other or two south ends near each other, they **repel** (push apart from each other). But if you bring a north end near a south end, they **attract** (pull toward each other). Magnets can also be objects containing iron, nickel, or cobalt. Bring the end of a magnet near a metal paper clip, and the magnet will attract the clip. Bring the magnet near a penny, and the magnet will have no effect. The paper clip is made of iron, but the penny is made of copper, which is not magnetic.

ANIMAL MAGNETISM. *Homing pigeons are pigeons that use the Earth's magnetism to **navigate** (find their way). In areas where there is a magnetic disturbance, and compasses don't point north, these homing pigeons might get very confused, and might not find their way home!*

Why does the needle of a compass point north? Metals deep in the Earth's core make it seem as if the Earth has a giant magnet in it. The needle of a compass is actually a little magnet. One end of the compass needle is attracted by the Earth's magnetic north pole and repelled by its magnetic south pole. For the other end of the compass needle, it's the other way around. That's why the compass needle lines up along a north-south line, showing you which way is north.

N *orth, South, East, or West? How to make a Chinese hanging compass.* The Chinese were the first to make use of magnetism over 900 years ago. To make one of their early compasses, rub one end of a magnet along a needle. *Be careful not to prick yourself with the needle.* Always rub in the same direction about 20 to 30 times. The needle will become magnetic. To test it, see if it will pick up a pin.

Tie one end of a short thread around the center of your magnetized needle. Tie the other end around a pencil. Place the pencil across the rim of a glass that's wide enough to let the needle swing freely when you lower it into the glass. When the needle comes to rest, it will point in a north-south direction.

What if the Earth spun very fast

You'd have to tie yourself down to the Earth with some very strong ropes or cables. If not, you might fly off into space!

Each day the Earth makes one complete spin on its axis. It spins so slowly that you don't even notice. Now let's suppose that the Earth spins much faster than it really does. To imagine, think about those spinning rides in some playgrounds. What happens when somebody gives the ride a hard spin? When the ride spins fast, you have to hang on or you'll fly off.

If the Earth spins just 17 times faster—17 complete spins in one day—that would be fast enough for objects (and people) to fly off at the **equator** (the imaginary line around the center of the Earth). Things would fly off at the equator first, because that's the farthest distance from the Earth's axis.

Let's say you're exploring right near the equator when the

Round and Round and Round She Goes *The effects of spin.* Find a turntable like a "lazy Susan" that you may have in your kitchen. Put some pennies close to the center—or axis—of the lazy Susan. Put some others near the edge. Before you give the lazy Susan a spin, try to guess which pennies will fly off first.

You probably found that the pennies farthest from the axis (or nearest to the edge) flew off first.

Earth starts spinning as fast as 17 times a day. You'd need to be tied down to keep from flying off into space. But people standing near the North Pole or South Pole, which are right on the Earth's axis, wouldn't notice anything different. (Actually, if the Earth really did spin that fast, the oceans and the air would be flung into space. People everywhere would be in big trouble, no matter where they lived!)

What if the Earth suddenly stopped spinning

Y ou'd feel like your heart was flying out of your mouth. If the Earth stopped spinning all of a sudden, objects would be flung eastward at a high speed. At the equator their speed would be 1,000 miles an hour (1,600 kph)!

How come? Objects near the equator travel all the way around the Earth—about 24,000 miles (40,000 km)—every 24 hours. So they are traveling 1,000 miles every hour—or a speed of 1,000 miles per hour (1,600 kph)—all the time, just because of the Earth's spin. These objects would still be moving at that same speed (in an easterly direction), even if the Earth were to suddenly stop spinning.

You can feel the same sort of thing happen if you are riding in

S *top the Earth, I Want to Get Off* *What happens to objects if the Earth stops spinning?* Put an empty soda can near the edge of a turntable like a lazy Susan. *Slowly* start making the lazy Susan spin faster and faster. Now, suddenly stop it from spinning. The can will probably topple forward.

Repeat the experiment, moving the can closer to and farther from the center. You should find that the farther the can is from the center, the more likely it is to topple. Before you do the experiment, see if you can guess what will happen if the can is right at the center when you suddenly stop the spin.

a car that suddenly stops. You feel like you are being thrown forward. But you aren't really *thrown* forward. You just keep moving, while the car comes to rest. That's why we wear seat belts—to keep us from moving forward when the car stops suddenly.

K**EEP ON TRUCKIN'.** The English scientist Sir Isaac Newton (1642–1727) described the way all moving things keep moving by themselves unless something else slows them down. We call this property of all objects inertia. Newton's law of inertia says that without any outside force, an object at rest stays at rest, and a moving object keeps moving at the same speed in a straight line.

It's the same with spinning objects. If there's nothing to slow an object down, it will just keep spinning forever all by itself. But you've probably noticed that spinning objects stop spinning, even if you don't stop them by hand. These objects are stopped by friction. Friction is the rubbing of one thing against another, such as air against objects. Raise the front tire of your bicycle off the ground, and give it a spin. Friction is what makes the tire eventually come to rest.

How about the spinning Earth—will it also eventually come to rest? There is a very small amount of friction caused by the air and ocean tides. The air and the water rub against the land and each other. This friction causes the Earth to slow down very gradually. The Earth spins on its axis once each day. So, the days will get slightly longer as the Earth's spin slows down. But the slowing down is so slight that the days will get longer by only a few seconds after a million years. ◀

What if the Earth were covered by water

I f the Earth were entirely covered by water—blub blub blub—you probably wouldn't be here reading this book—gulp! Or if you were, you'd be one very smart fish.

THE STUFF OF LIFE. *Earth is the only planet in the solar system with liquid water on its surface. Europa, one of Jupiter's moons, may have an ocean of water under its surface.*

Scientists believe that life on Earth first began with tiny **organisms** (living things) in the oceans. These organisms **evolved** (developed) into more complicated life forms that eventually crawled onto the land. But with no land at all, land animals—including people—wouldn't have evolved.

After all, why have legs if there's nothing solid to walk on?

Today, over two-thirds of the Earth's surface is covered by water in lakes, rivers, seas, and oceans. Turn a globe of the Earth just the right way, and you hardly see any land at all on one side. Why isn't the Earth covered entirely by water? The dry parts are there because the Earth is not perfectly smooth and round.

What if it didn't stop raining?

Wouldn't the water level rise and cover the land? Rain does make the water level rise, which sometimes causes floods. But where does rain come from? It comes from **evaporation**—water leaving the surface of lakes and oceans and going into the air in the form of a gas or vapor. So rain doesn't add to the total amount of water on the Earth. It just moves the water around from one place to another. That movement of water is called the **water cycle.**

If it were, all parts of the Earth would be covered by a layer of hundreds of feet (meters) of water. But because the Earth is not perfectly smooth and round, some slightly raised parts of the sphere stick up above the water and form the land.

"OK, there's nothing but water everywhere, so what should we evolve into?"

If the amount of water on Earth were twice as much as now, everything would be under water. With somewhat less water than that, only the tallest mountains would remain uncovered. These would become small islands surrounded by water. But don't worry, it's impossible that this could actually happen, no matter how much it rains.

What if the Earth warmed up and all the ice near the North Pole and South Pole melted?

Even if this happened, the oceans would rise by only about 30 feet (just over 9 meters). Such a rise in sea level would cause big problems if you lived near the ocean. Coastal cities such as New York and San Francisco would be under water. But it wouldn't change the total fraction of land under water by very much, and it's not likely to happen.

It's Raining Frogs
Weather and Climate

The weather changes from day to day, and **climate** is the average weather in a certain place. In deserts, for example, the climate is very dry, even though it may sometimes rain. Some places on Earth—particularly places near the Earth's equator—have almost the same climate all year round. As you go closer to the North Pole or South Pole, the difference between seasons becomes more extreme.

Changes in temperature during the year are caused by the warming effect of the Sun's rays. Sunlight warms the Earth much more in summertime when the Sun is high in the sky than during winter when it's low in the sky.

The warmth of the Sun also causes air to expand and move and create the winds. The big storms called hurricanes, cyclones, and tornadoes happen when air masses start whirling in a circle around a region of low pressure. Storms and other changes in the weather can be tracked by satellites. Scientists can also use satellite measurements and computers to predict reasonably well what the weather is going to be in a few days. It's much harder to predict changes in the Earth's climate over a longer period of time. Most changes in the Earth's climate happen slowly. But some sudden changes in climate could happen if, for example, the Earth were hit by a giant meteorite.

What if the Earth were hit by a big meteorite

A meteorite is a rock from outer space that hits the Earth. If a big meteorite hit the Earth, chances are that you wouldn't see it, but you might feel its effects. A giant meteorite would probably land in an ocean, because over two-thirds of the Earth's surface is covered by water. When it hits the water, the meteorite would create a giant wave.

But a rock from outer space heading our way might not even reach the Earth's surface. The Earth is surrounded by a layer of gases called the **atmosphere.** A meteor that passes through the Earth's atmosphere heats up because of friction. Small meteors heat up so much that they burn up before they reach the ground. These are called shooting stars. On some nights of the year, you can see shooting stars moving quickly across the sky.

A giant meteorite striking the Earth might also create a very big explosion. Yet, even a giant meteorite wouldn't knock the Earth out of its orbit. The Earth is simply too big. For example, suppose a meteorite 10 miles (16 km) wide hit the Earth. The diameter of the Earth is about 800 times bigger than that. It would be like a tiny speck of dust hitting a basketball.

WHAT DO YOU CALL THOSE ROCKS FROM OUTER SPACE? *Scientists have three classifications for rocks from outer space:* **meteoroids** *when they are in space,* **meteors** *when they enter our atmosphere and burn up before landing (these are shooting stars), and* **meteorites** *when they don't burn up in the sky but land on Earth.*

But a giant meteorite could affect the Earth's climate. If a huge meteorite hit the surface of the Earth, a lot of dust would be tossed up into the atmosphere. All this dust would be thrown up so high that it would stay in the atmosphere for a long time. Lots of dust in the atmosphere blocks sunlight and makes the Earth cooler.

If the meteorite were a very big one, the sunlight might stay blocked for as long as a couple of years. This would be very bad news for most plants, animals, and people. People could stay inside to get out of the cold. But plants wouldn't be able to grow without sunlight, and food would become very scarce. Scientists think that a big meteorite hitting the Earth may be what killed the dinosaurs and many other animals 65 million years ago.

Could it happen today? Every once in a while we find a big meteorite that just whizzes past the Earth. Based on the number of "near misses," scientists think that a big meteorite might hit us once every 5,000 years or so. But that means the chances of a meteorite hitting Earth is one chance in 5,000 every year. Some scientists think that we should keep a lookout for big meteorites heading this way. If we spotted one heading toward us while it's still far away, we might be able to knock it off course with a big bomb.

"Dull, dull, dull. *Nothing* ever happens around here."

What if a day lasted a year

You'd be either fried or frozen, depending on where you lived. The Earth spins on its axis once each day, and it takes the Earth 365 days to complete its orbit (path) around the Sun—that's why there are 365 days in a year. But imagine that instead of spinning 365 times, the Earth spun only once on its axis as it orbited the Sun. In that case, each day would be a year long.

If a day lasted a year, the Earth would always keep the same side facing the Sun. (To see why this would happen, see the Longest Day experiment on the next page.) One side of the Earth would always be in sunlight, and one side would always be in darkness.

The Longest Day *How a day can last a year.* Put a small lamp in the middle of your kitchen table. This lamp will represent the Sun. Take a globe of the Earth, or even a ball, and walk it in a large circle around the Sun. As you make the Earth orbit the Sun, slowly turn the Earth so that the same side always faces the Sun. You will find that the Earth turns exactly once on its axis exactly when you make a complete orbit around the Sun.

This would have a drastic effect on the Earth's climate. The "sunny-side-up" side of the Earth would be extremely hot, and the dark side would be extremely cold.

People on both sides of the Earth would probably pack their bags and head off to live in the narrow boundary between the two sides. This boundary would be a thin circle around the Earth. It would be the only place where it would be possible to live without freezing or burning up.

The boundary region would also be the only place that water would exist. The bright side of the Earth would probably be so hot that any water there would evaporate (turn to gas) and go into the air. High winds would carry the moist air to the dark cold side where the water would be frozen out of the air and come down as snow.

Could a day ever last a year? The spin of the Earth is gradually slowing down. But it will never slow down enough for the Earth to always keep the same face toward the Sun. Instead, the Earth's spin will probably eventually slow down to one spin every month (and keep the same face toward the Moon)—but not until millions of years from now.

What if there were no seasons

You might not know when it was time to start summer vacation! If there were no seasons, no matter where you lived it would be about the same temperature all year round, just as it is now around Earth's equator. If the average temperature were the same as now, winters would be warmer than now and summers would be cooler. Plants could grow just as well at any time of year. And birds wouldn't need to migrate (travel to warmer places during the winter).

Why do we have seasons? Some people think that seasons exist because the Earth doesn't move around the Sun in a perfect circle. Sometimes the Earth's a little closer to the Sun, and sometimes it's a little farther from it. The Earth really does change its distance to the Sun during the year, but that's not why we have seasons.

We have seasons because the Earth's axis is tilted. Summer occurs in the Northern **Hemisphere** (half of the Earth) when that part of the Earth is tilted toward the Sun. The Sun appears high in the sky then. Its rays shine almost straight down, so they warm the Earth a lot.

People living below the equator live in the Southern

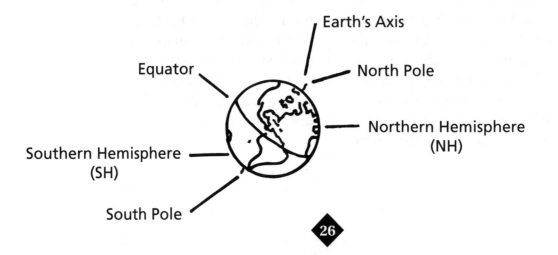

Earth's Axis

Equator

North Pole

Northern Hemisphere (NH)

Southern Hemisphere (SH)

South Pole

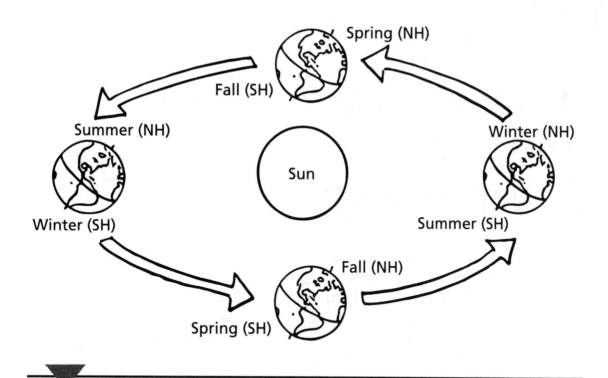

Spring (NH)

Fall (SH)

Summer (NH)

Winter (SH)

Sun

Winter (NH)

Summer (SH)

Fall (NH)

Spring (SH)

ENDLESS **S**UMMER. Near the equator the sun is high in the sky all year round, and it's always hot. Instead of having spring, summer, fall, and winter, places near the equator have rainy and dry seasons. Rainy seasons occur when winds carrying moisture blow in from the ocean. Dry seasons occur when winds blow in from over land. ◀

Hemisphere. They have winter when the people in the Northern Hemisphere have summer. In winter the Sun is low in the sky, and its rays hit the Earth at a small angle. Because of this angle, the rays don't warm the Earth as much as they do in summer. Six months later, when the Earth is on the opposite side of its orbit, the Northern

Hemisphere has winter and the Southern Hemisphere has summer.

If the Earth's axis weren't tilted, there would be no seasons. The Sun's rays would warm the Earth the same amount no matter where it was in its orbit around the Sun, and the weather where you live would never change very much at all.

What if it rained cats and dogs (or fish and frogs)

You'd need a very strong umbrella to keep furry creatures from landing on your head. People sometimes say that "it's raining cats and dogs" when it rains really hard. This expression is not as crazy as it sounds. Some people (and not just people in fairy-tale books, either) have seen it raining fish or frogs.

How could this happen? Here's one possible explanation: **tornadoes.** Tornadoes, or "twisters," are violent, spinning, funnel-shaped winds that can lift cars and even whole houses up into the air. A tornado occurring over a lake could lift a lot of water—and the fish and frogs in it—up into the sky. Then, plop! The tornado might drop the fish and frogs over some other place, miles away, and it would be "raining" fish.

Other surprising things sometimes fall from the sky. Hailstones are balls of ice the size of peas or bigger. **Hail** happens when falling raindrops freeze. Raindrops never get very big, but hailstones can be the size of grapefruits. This happens when ice crystals fall through the atmosphere, and many layers of frozen water are added during their fall to Earth. Don't go out into grapefruit-size hail even if you have a crash helmet!

And if getting hit by giant ice balls isn't bad enough, watch out for meteorites. At first, scientists didn't believe that rocks fell from the sky. They thought that people who told such stories were crazy. Now we know that the stories are true. There are lots of rocks in space, left over from the time when our solar system formed billions of years ago.

What if people could control the weather

HOME IN A DOME: BIOSPHERE 2.   It's possible to control the weather inside an enclosed area by building a large dome over it. In the middle of the Arizona desert is an enclosed dome called Biosphere 2. It was built in 1991 in order to see if humans could create a copy of Earth's **biosphere**—its atmosphere and **ecosystems**—in which they could live without supplies from the outside world.

Biosphere 2 contains miniature versions of Earth's five ecosystems—desert, grassland, marsh, ocean, and rain forest. It even has a coral reef. Eight men and women stayed inside Biosphere 2 for two years. The experiment didn't work too well—the dome wasn't completely sealed off from the outside. Now the dome is used to study the environment and the effect we humans have on it. But, if we could perfect a sealed environment in which people could live, then humans could live in Earth-like biospheres on the Moon or even on Mars. ◄

It would never "rain on your parade." You'd never have to cancel a ski trip because there wasn't any snow. And it would never get so hot in the summer that you'd feel like your skin was melting! But, wait a minute, what if your sister liked it that hot? Or your neighbor wanted rain all the time? This could get complicated!

Scientists can't control the weather, but they can sometimes create rain when there are a lot of clouds. Clouds are just large collections of tiny water droplets suspended in the sky. Rain normally occurs when the tiny droplets in a cloud combine and make bigger droplets. The bigger droplets then fall out of the sky as rain. Scientists can make rain by putting certain chemicals inside clouds. The chemicals cause the water droplets in the clouds to combine and form bigger droplets that will fall out of the sky. But this method is difficult and expensive to do over a very large area.

Besides making rain, scientists might also want to *stop* rain

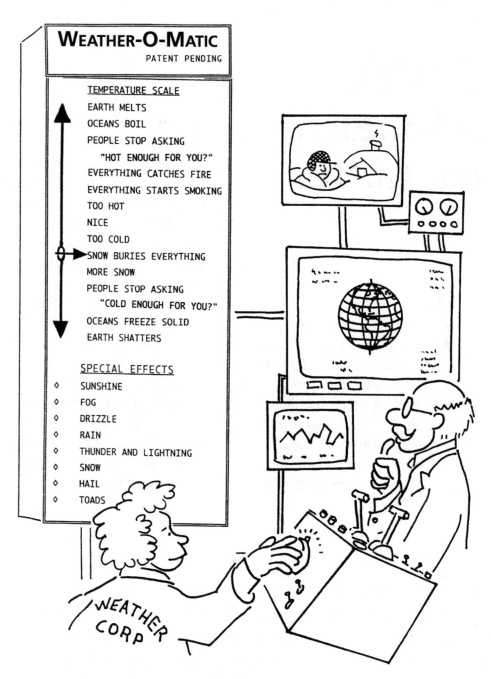

and prevent flooding, or to keep big storms from happening. Hurricanes are very big storms that start over the ocean when a large region of air begins whirling around in a circle. So far, we don't know enough about exactly how hurricanes start to be able to tell just where and when they will occur.

What if the climate got much hotter

If all parts of the Earth warmed up very slowly over thousands of years, plants and animals would be able to adjust to the change. But what if the Earth warmed up quickly?

If the Earth warmed up a lot in the next 100 years, some species of plants and animals might be able to move to cooler regions of the Earth. Some others might die out.

Many scientists worry that the Earth's climate is warming up too fast, and that we need to do something before it's too late. But it's hard to tell exactly how much the whole Earth has warmed up. Temperatures differ from one place to another and throughout the year. If the Earth is warming up, it hasn't warmed up very much so far.

Why would the Earth be warming up? As sunlight warms the Earth's surface, the heat can't always escape. It's trapped in the atmosphere by certain gases such as carbon dioxide. CO_2 is

GET ME OUTTA HERE! How heat gets trapped. *The next time you get in your family's car on a warm sunny day, notice how hot the car is when you first get into it. How did the inside get so hot? The sunlight comes through the windows to warm the inside, but the heat can't get out through the glass windows. The inside of the car gets hotter and hotter until someone opens a door or window to let the heat out. For this reason, it's very dangerous for people or animals to be in a parked car with the windows closed. The Sun will heat the inside of the car whether the day is warm or cool, even on a cloudy day.*

the chemical name for **carbon dioxide,** which acts like a blanket that keeps the Earth warm at night. Some of this warming is good. If it didn't happen at all, the temperature of the Earth's surface would be below freezing, and we couldn't live here comfortably.

But more and more CO_2 is created by burning fuels such as coal, oil, and gas. These fuels are burned to make electricity and to produce heat. The amount of carbon dioxide added to the atmosphere keeps increasing— it's like putting more and more blankets on the Earth—and the Earth could get overheated.

What if the sky were always cloudy

If the sky were always cloudy, less sunlight would reach the Earth's surface, and it would rain more. With little sunshine and more rain, the kinds of plants that grow on Earth would be different, and that would affect all of the life on Earth. But another important effect might be that we'd know less about the world. If the sky were always covered with thick clouds, we might not know about the stars, or even about the Sun and the Moon. We might not even realize that the Earth was round!

A long time ago, people thought that the Earth was flat. Eratosthenes, a man who lived in Greece more than 2,000 years ago, found a way to measure the Earth's diameter and to show that the Earth was round. He did this by seeing that the length of a shadow of a pole was different at different points on the Earth. To get the idea, imagine that you are standing someplace on Earth where the Sun is directly over-head. You have no shadow at all. Now imagine two friends stand-ing far away at other places on the Earth. Because the Earth is round, your friends do have shadows. But if the sky were always cloudy, none of you would have a shadow, and the experiment wouldn't work. (See the Shadow Play experiment on the next page.)

If the sky were always cloudy, people might not have explored the Earth. In the early days of sailing ships, sailors watched the positions of the stars to tell what direction the ship was traveling.

Could Life Survive?

You might think that life couldn't survive if the days were always cloudy, but that's not necessarily so. Clouds block sunlight and usually cool the Earth, but they can also act like a blanket to keep heat from escaping. If the cloud layer wasn't very thick, temperatures on Earth might not be much lower than now, and they might even be higher. And enough light would reach the Earth so that many plants could grow.

Without being able to see stars, sailors probably wouldn't have sailed very far from land. They might have worried about finding their way back, or even about falling off the "edge" of the Earth if they thought it was flat.

And forget about traveling to space. If it were always cloudy, we wouldn't even know that the Moon and stars were there, so there'd be no reason to send a rocket up to see them. No astronaut would have come back with pictures of our big, blue and white, totally round Earth!

Shadow Play *Using shadows to see if the world is round.* Get some toy soldiers all the same height. Tape the soldiers standing at several different points around a globe of the Earth (or a ball). Place a lamp at the middle of your kitchen table. Hold the globe some distance away. Notice how the length of each soldier's shadow is different depending on where it is on the globe.

Now tape some toy soldiers standing on one side of a flat cereal box. Hold the cereal box away from the lamp. Their shadows should all be about the same length. If we lived on a flat Earth, the length of our shadows wouldn't change as we moved around the Earth.

FROSTED SUGAROOS

FROSTED SUGAROOS

Ingredients:

sugar, sugar, more sugar, an oat

Nutrition Value:

What's the Matter with Matter?

Forces and Energy

Matter is the stuff that everything is made of. If you divide a piece of matter into smaller and smaller pieces, you'd eventually find a smallest piece that keeps the properties of the original matter (called a **molecule**). If you split up a molecule, you'd get **atoms,** the building blocks of all matter.

The three types of matter are called solids, liquids, and gases. **Solids** either have a fixed shape or tend to keep their shape—like a piece of clay. **Liquids** take the shape of their container, and they are **fluid** (they can flow). Both liquids and solids occupy a fixed **volume** (amount of space). So, even if you squeeze a solid or liquid (put it under pressure), you won't change its volume very much. **Gases** are also fluids, but they don't occupy a fixed volume. That's because the molecules in gases are not bound closely together like they are in liquids and solids. The molecules in a gas can spread out and fly apart. That's why gases are so much lighter than solids and liquids.

All three forms of matter can have **energy,** which is defined as the ability to do work. Energy cannot be created or destroyed. It can only be changed from one form to another. The energy of moving objects is called **kinetic energy,** and the energy something has because of its position or condition is called **potential energy.** Water in a waterfall loses potential energy and gains kinetic energy as it falls. Other forms of energy include heat, electricity, and light. Most of the energy on Earth originally comes from the light from the Sun.

What if things fell up, not down

You'd have a hard time picking up anything you dropped! Keys, money, balls, or candy would fly up into space. Maybe people would carry butterfly nets to catch things they dropped. But there's a fun side of "antigravity" (things falling up not down)—you could use an ordinary rug to make a magic carpet ride into space.

Long before people actually went to the Moon, the writer Jules Verne (1828–1905) wrote a book called *From the Earth to the Moon,* in which he described using an imaginary antigravity material to make a trip to the Moon. Nobody thinks that such a material could really exist. If it did exist, it would fly off into space by itself unless it was somehow held down.

But wait a second! There is a material that falls up. It's a gas that we call **helium.** Of course, a balloon filled with helium does not rise because of antigravity. So, why does it rise? Remember that air is a gas, and we live at the bottom of an "ocean" of air called the atmosphere. In the real ocean, materials that are lighter than water (such as wood) will rise rather than sink. In just the same way, helium balloons rise because helium gas is lighter than air. Helium balloons can even lift something tied to

Fun with a Helium Balloon *How much heavier is air than helium?* Get a balloon filled with helium. Tie a 3-foot (1-m) piece of thread to the balloon. Next, tie a paper clip to the thread every 2 inches (5 cm) along the length of the thread. Let the balloon go, and see how many paper clips it can just barely lift off the ground. The weight of those paper clips is how much heavier the air is than the helium balloon. A bigger helium balloon would lift more paper clips.

them—
as long as
the object plus
the helium weigh
less than the surround-
ing air. A giant helium
balloon could even lift you.

What if you tried to take a space trip using giant helium balloons?

Right now it takes a lot of energy and money to launch rockets into space. You might think space trips would be free if we could hang spaceships from giant helium balloons. The ships would rise all by themselves, right? Unfortunately, using helium balloons for space trips is not possible. The balloon can rise only as long as both it and its attached spaceship are lighter than the surrounding air.

Think about the ocean again. A piece of wood can rise only to the water surface. It can't rise out of the water. The air in the atmosphere becomes thinner and lighter, the higher up you go. If you go high enough, you run out of air. So a helium balloon (even with no spaceship hanging from it) could rise in the sky only as long as the air is heavier than the gas in the balloon—only a few miles up. Helium balloons certainly couldn't carry things into space, where there's no air.

What if things could pass through one another

The coolest part would be that you could walk through walls, like a ghost. If your parents said, "You're not leaving that room until you do your homework," schloop, you could walk right through the closed door and out of the house! The not-so-cool part would be trying to pick something up only to have it pass right through your hand.

In real life, solid objects have definite shapes that don't change easily. When one solid object passes through another, it leaves a hole in the object it passes through. If you hammer a nail into a piece of wood, and then pull out the nail, you'll find a hole in the wood.

Most people don't believe in ghosts. But there is something ghostly that can pass through a wall without leaving a hole. It's called **ball lightning.** This is a very unusual type of lightning that appears in the form of a glowing ball. Ball lightning also moves much more slowly than the lightning you usually see during a storm. Unlike normal lightning, ball lightning does not damage objects that it passes through. Some people have seen ball lightning pass through the walls of houses and airplanes without leaving any holes. Scientists aren't sure how ball lightning works, and they can't predict when it will occur.

Look Ma, No Holes *See a wire passing through a block of ice.* Fill a square pan with water and put it in your freezer. When the water is completely frozen, take the ice out of the pan in one piece. (If you put the pan in warm water, the ice will come out easily.) Place two kitchen chairs back to back. Rest the block of ice on the backs of the chairs. (Make sure you are allowed to put a block of ice on the chairs!) Get a long piece of fine metal wire, and tie a hammer to each end of the wire. Place the middle of the wire on top of the block of ice, with the hammers hanging off both sides.

Wait a while and you will see the wire will move down through the ice without cutting it. The wire moves down because the force of the weighted wire melts the ice under it. *Voila!* The wire doesn't leave a hole, because the ice refreezes after the wire moves through it. The same kind of thing happens when you ice-skate. The blades of the skates press down hard enough to melt the ice in a groove, which then refreezes.

What if water didn't evaporate

Get your fins on, because if water didn't evaporate, we'd be living in the ocean! All the water would be on the surface of the Earth, and none would be in the air, so it would never rain. Rain can happen only when water evaporates from lakes and oceans. Rain is needed to grow plants and to clean the air. With no rain, streams and rivers would just dry up. So, if water didn't evaporate, the only life on Earth would probably be in large lakes or oceans.

Evaporation is the process of a liquid (for example, water) turning into a **vapor** (gas form of a liquid). Water, like all other substances, is made up of tiny particles called molecules. In liquid form, water molecules move every which way, sort of like dancers, but the dancing molecules are almost touching their neighbors because of strong forces attracting them together. When water is heated, however, its molecules move faster and faster, like dancers at a rock concert. Eventually, some move fast enough to break free and go into the air as vapor.

SWEAT MAY BE SMELLY, BUT IT'S COOL. *If water couldn't evaporate, we would not be able to cool off by sweating. Sweat may smell yucky, but it's important. When you sweat, the fastest molecules in sweat break free during evaporation, taking heat energy with them.*

To picture evaporation, just imagine that you are holding a bunch of puppies, each one on a separate leash. The most excited puppy—the one who is running back and forth the fastest—will probably be the first one to pull the leash out of your hands and break free. In much the same way, the fastest-moving water molecules are the ones that can break free of the other water molecules and evaporate.

SNAP!

What if drops of water were really big

Imagine raindrops the size of water balloons. You'd get completely soaked if you were hit by just a single raindrop. You might even get hurt.

Could raindrops ever get to be as big as water balloons? Rain starts out as very tiny water droplets in clouds. As the droplets fall, they hit other droplets and get bigger. Big raindrops have hit a lot of other small raindrops on the way down. But raindrops never get to be as big as water balloons. Once a raindrop gets to be a certain size, the air rushing past as the drop falls tends to break it apart.

Water is made up of tiny moving particles called molecules that are attracted to each other. Raindrops or any other drops of water could get as big as water balloons only if water molecules were attracted to each

NEED A DRINK? *Sucking big drops of water through a straw would be nearly impossible even if you had a really big straw. Imagine trying to suck a water balloon through a toilet paper tube!*

other with much more force than actually exists. Let's pretend that that much force exists. You might not want to turn on your faucet! Taking a shower would be a challenge. You'd keep getting bonked on the head by huge drops. And such big drops of water would not be so easy to get off with a towel.

Water Works *How to keep water in a straw.* Stick a straw in a glassful of water. Take the straw out of the water, and water falls out the bottom of the straw. When the water falls out, air is coming in the top of the straw to take the water's place. Now, put the straw in the water, hold the straw in your hand, and cover the top of the straw with your thumb. Lift the straw out of the water and the water should stay in the straw—at least until you take your thumb off the top. Water stays in the straw because the air can't come in to take the water's place. The water won't come out if the air can't get in. The forces between water molecules keep them together, and they block the air from coming in the bottom of the straw. If the force between water molecules was much stronger than it actually is, more water molecules would stay together. And the water would not fall out even if you did this experiment using a straw much wider than a normal one. You could even turn a glass of water upside down, and the water wouldn't fall out!

Step 1

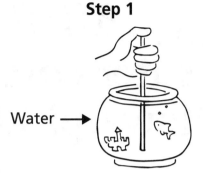

Water →

Step 2

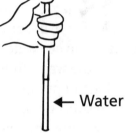

← Water

Step 3

Water →

Step 4

← Wet Cat

45

What if water could flow uphill?

"But it does," you might say. "Anybody getting a drink out of a water fountain can see water flowing up!" But the water fountain uses some kind of pump to get the water to spurt up. By itself, water only flows downhill, not uphill, because of the pull of gravity.

Water that flows over a waterfall was formed as a result of evaporation and then became rain and snow, which in turn created rivers and streams.

A *Flowing Rope.* Get a piece of thick rope or cord such as a clothesline. Put most of the rope on your kitchen table, and allow part of the rope to lie on the floor. Place a pencil or pen under the middle of the rope. Raise the pencil and you'll be raising that portion of the rope above the table. The rope should rise off the table and "flow" over the pen until it all winds up on the floor.

When water flows downhill in a waterfall, the falling water has a lot of energy. If you make the water turn some wheels as it falls, you can turn that energy of motion into another form of energy—electricity. Let's suppose that the water at the bottom of a waterfall somehow could flow back up to the top by itself. If that could happen, you'd have a way of making any amount of electricity for free, without having to wait for rain or snow to get the water back up to the top of the falls.

To make water from a waterfall flow up to the top, you'd need to pump it back to the top. If you had a perfect pump, the amount of electricity used to pump the water back up to the top would be the same as the electricity produced by the falling water. But perfect pumps don't exist. Some energy is always changed into heat and wasted. In order to get the water back up to the top of the waterfall, a real pump would need *more* energy than the electricity that the falling water could produce.

You could also get water to go uphill using something called

U pside Down Waterfalls
How to make a siphon.
You can make a siphon using a tube of bendable plastic or rubber. Put one end of the tube in a glassful of water, and put an empty glass next to the full glass. Make sure the full glass sits higher than the empty one. Suck on the other end of the tube a little to start the water flowing. Then quickly cover that end with your thumb, before any air gets in. Put this end in the empty glass. The water should flow through the siphon until the water level is the same height in both glasses.

Step 1

Step 2

Step 3

Step 4

a **siphon.** A siphon is a tube filled with water with each end placed in separate containers of water. If the water level is higher in one of the containers, water flows up the tube from the container with the higher level and back down the tube into the other container.

How does a siphon suck the water up? You can imagine that the water in the tube is like a rope. The end of the siphon that is in the container with a lower water level has a longer length sticking out above the water, and it has more water in it. So, the water rope is heavier on that side. The heavier side falls, and pulls the water "rope" up the other side of the tube. The forces between water molecules keep them all together and keep the water rope from breaking.

Are You Seeing Things?
Light and Sound

Light and sound help us get to know our world. You see objects when light bounces (**reflects**) off the object and reaches your eyes, or when an object gives off its own light. In a dark room you can't see any objects that reflect light, but you can see a glowing object like a lightbulb. In order to hear something, sound must reach our ears. Sound, like light, can reflect off objects. We call reflections of sound **echoes.**

Both light and sound are forms of energy, and both travel in waves. Picture some water waves created when you drop a pebble in a pond. They look like a bunch of expanding circles. The high point on each circle is called the **crest** of a wave. And the distance from one wave crest to the next is called the **wavelength.** Sound waves that have a short wavelength are heard as having a high **pitch** (squeaky sounds), and sound waves with a long wavelength are heard as a low pitch (deep sounds, such as thunder). Some sound waves with very short or very long wavelengths cannot be heard by humans. Very short and very long wavelength light waves cannot be perceived by the human eye.

Waves too long to be seen include **infrared** (sometimes called heat waves), **microwaves,** and **radio waves.** Waves too short to be seen include **X-rays** and **ultraviolet.** These waves are all around us in the environment. Even though we can't see these waves directly, we can use their energy.

What if everything were the same color?

Check it out—try fiddling with the color controls on your TV. Watch one of your favorite programs with everything in shades of green, red, or blue. Do you lose interest in the show, or do you get used to the one-color world?

Changing to a one-color world on TV is one thing. In the real world, not only would you get bored, but Mother Nature would be disturbed. Animals and plants use color to attract mates, to find food, and sometimes to warn others away. In a one-color world, animals and plants would need other ways to do these things.

But how could everything be seen as one color? The color of an object depends partly on the light that shines on it. If only one color light, say red, shone on objects, everything would look different shades of red. So, if one color light from the Sun reached the Earth, everything on Earth would look that color.

You're probably thinking that only one color light from the Sun *does* reach Earth—sunlight isn't all different colors. Well, actually it is. The white light is a mixture of all the colors of the rainbow. (Check it out in the Reversing the Rainbow experiment.) Light travels in waves, and each color of light has a different wavelength. Each wavelength is seen as a different color. Objects look all different colors because they reflect different amounts of different light waves. Blue objects reflect mainly blue light waves, and red objects reflect mainly red light waves. What about objects that reflect the same amount of all colors? They'd look white if they reflect a lot of each color, or gray if they reflect less of each color.

Reversing the Rainbow *How to mix colors and get gray.*   Draw a circle about 2 inches in diameter on an index card. Use a pencil to divide the circle like a pie into 6 slices. Paint or use markers to color one section each red, orange, yellow, green, blue, and purple. Use a sharp pencil to punch a hole in the center of the card. Be careful with the sharp point! Push the card down to the middle of the pencil. Now spin the card very fast by hitting its edge with one finger. All the colors should blur together to make a grayish color.

What if the sky were black during the daytime

Get out your space suit. If the sky were black, you'd need to wear a space suit in order to breathe. The sky is black at night, but a black sky during daylight happens only when a planet has no atmosphere—like the Moon. Earth's sky appears blue, because of the light coming through the gases making up the air in our atmosphere.

OXYGEN

SCHOOL BUS

STOP ON SIGNAL

NO AIR

834

The sky might be a different color. For example, pictures taken by spacecraft on **Mars** show a pinkish orange sky. The atmosphere on Mars is mostly carbon dioxide, but it's actually the very fine red dust in the atmosphere that gives the Martian sky its color. But you don't have to go to Mars to see an orange sky. Fine dust in the air can sometimes make the sky pink, orange, and purple on Earth, too, during sunsets.

The Earth's atmosphere is mainly a mixture of two gases—the **oxygen** we breathe, and **nitrogen.** We usually think of air as having no color at all. After all, we can see right through it. But if sunlight bounces off a very large amount of air, the color of light that reaches our eyes will be blue.

Have you ever seen mountains from very far away? They probably appeared bluish, too. That's because what you're actually seeing is the air between your eye and the faraway mountains. The farther away the mountain is, the bluer it looks because you're looking through more air.

The color of the sky is not always blue. It depends on what's in the atmosphere besides air. In places where there's a lot of pollution, the sky may appear more white or brownish than blue.

What if light didn't travel in straight lines?

Y ou'd be able to see things happen behind your back. If someone was sneaking up on you, you'd know it in advance. You could turn around and scare the wits out of them!

If light didn't travel in straight lines, it could bend around objects. The light from objects behind you could bend around you and reach your eyes. Think about it—that's exactly how sound travels. You can hear music from a radio behind you or in another room. Sound doesn't have to travel in straight lines to reach your ears.

Why does light travel in straight lines, but not sound?

Both light and sound travel in waves. Both spread out from the place where they're created. Throw a pebble into a pond, and you'll see expanding circles coming from the point where the pebble entered the water. Both light and sound spread out that way. If the expanding water waves run into a small stick in the water, they'll bend around it. That's how sound waves behave. But if the expanding water waves run into a large object, they'll be blocked by it and won't bend around it. That's how light waves behave. Light waves usually don't bend around things, because most obstacles that light waves encounter are larger than their wavelength.

There is one way to make light waves bend, and you've probably done it yourself, lots of times. Have you put a straw in a glass of water? It looks like the part of the straw above the water is not connected to the part of the straw below the water. That

Refraction in Action *How to bend light.* Put a straw, a pencil, or any long thin object into a clear glass. Pour water in the glass until it's half-way full. Now put the glass on a table, and look at the glass from the same level as the table. It should look as if your straw or pencil is cut in half and that the pieces don't quite match up. Refraction is in action here and at anytime you use glasses, a camera, or a magnifying glass. Their lenses change the direction of the light.

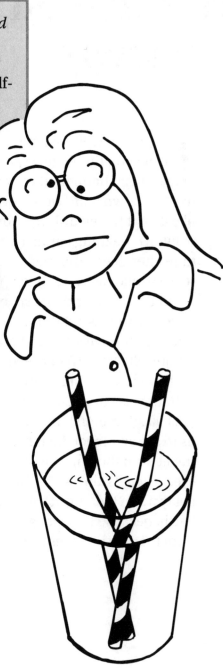

illusion is caused by refraction. **Refraction** is the change in direction of light when it passes from one medium to another, such as from air into water or glass. Light travels more slowly through water and glass than through air, so it bends as it enters them. (Check it out in the Refraction in Action experiment.)

What if light traveled very slowly?

Imagine that every time you turned on the TV news, you saw what happened yesterday. That's what life would be like if light traveled slowly. You wouldn't see things until a long time after they happened. You really would be living in the past.

In the real world, the speed of light is the fastest thing around—it travels 186,210 miles (about 300,000 km) per second. In the time you blink your eye, light travels about 20,000 miles (32,000 kilometers). In order for us to see an object, light has to travel from the object to our eyes. There is almost no time delay when looking at things on Earth, because the light goes from them to our eyes so quickly.

But say you're looking through a giant telescope at a star that's very far away. You're really seeing the star as it was years ago. Light *does* take a long time to reach us when we look at objects far out in space. In that case, we really are seeing things that happened a long time ago.

Okay, so light travels fast, but could it ever slow down? Remember the young boy mentioned in the introduction to this book who once thought about chasing a light beam? According to that young boy, we know that the speed of light in space never changes. That young boy grew up to be the scientist Albert Einstein. He came up with a world-shaking idea called the theory of **relativity.** His theory said that the speed of light always remains the same when it travels through space—but that time and length *do* change as objects travel very fast past us.

What if you saw a car traveling at almost the speed of light?

According to Einstein, the car would look as though it had been shortened. Also, everything inside the car would seem to be happening in slow motion. If the car tried to speed up a little bit, its length would shrink almost to nothing, and time would come almost to a standstill. But these changes only appear to someone watching the car go past. For somebody inside the car, everything seems normal. Strangest of all, the car just could not go any faster than the speed of light, even if it had the most powerful rocket engine imaginable. Einstein said that nothing can go faster than the speed of light.

What if you could see sounds

Your head would be crowded with pictures—the pictures of objects you see and of the sounds you hear. We can't actually see sounds with our eyes, but we *can* make pictures of sound.

FINDING A MEAL IN THE DARK. *Bats can echolocate so well that they can fly in complete darkness, avoid something as thin as a human hair, and locate something as tiny as a tasty insect.*

Doctors use sound waves called **ultrasound** to take pictures of a baby still inside its mother's body. Ultrasound waves have a very short wavelength. They can't be heard by our ears. The waves travel through the mother's body and bounce off the baby back to a machine that translates the sound waves into a picture. This is similar to the way the light rays called X rays can be used to take a picture of your bones or teeth. (See the X Facts for more on X rays.)

THE X FACTS. *X rays are a type of light wave, but they have a wavelength too short to be seen. Unlike visible light, X rays can easily pass through the skin of your body. But X rays cannot pass through your bones or your teeth—these parts are just too solid.*

Bats and dolphins are more talented "sound artists" than humans are. Bats send out a steady stream of ultrasound squeaks that humans can't even hear. Dolphins send out a series of clicks (sounding like rusty door hinges) into the dark sea. The sound of the squeaks or clicks bounces off objects and bounces back to the bat or dol-

58

Be Like a Bat and Do Like a **Dolphin**   *Using echoes to judge distance.*   Stand some distance away from the wall of a building. Clap your hands once, and listen for the echo. Walk farther away, and try again. You should find that the farther away the wall is, the longer it takes for you to hear the echo.

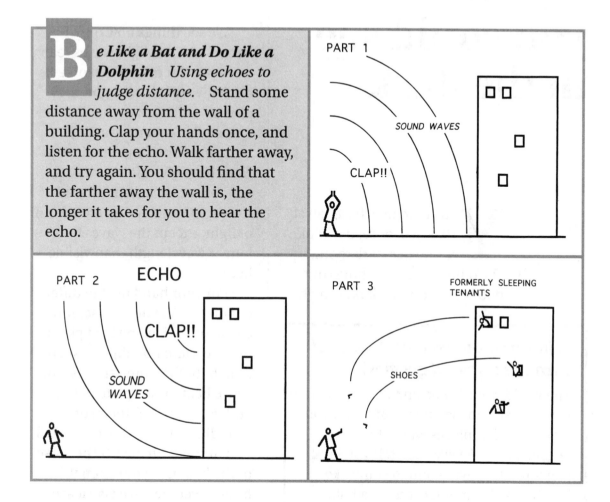

phin as an echo. Based on how long it takes for the echo to come back, they can figure out how far away the object reflecting the sound must be. The process is called **echolocation.** Echolocation not only gives an object's distance, but it also allows bats to figure out an object's size and shape. No one knows what goes on inside the brain of a bat or a dolphin. But scientists think that bats and dolphins get a picture of their environment from the echoes of their sounds. What do you think these sound pictures would look like?

What if you could see in the dark?

Y ou could play baseball at 10:00 P.M. or read a book sitting in a dark room. You'd be able to see things outdoors even on the darkest night.

What if you looked at the world through infrared goggles?

The world would look kind of spooky. If you look at a house outside at night, you would see that places where heat is leaking out look very bright. If you look at people, they will seem to shine like lightbulbs. You might even see a halo around them from the heat they give off. But if they wear glasses, the glasses look black, because they're cooler than the person's face. If you dip your hand in cold water, your cold hand looks black. If you put ice cubes in a glass of hot water, the ice cubes look like lumps of coal, but the hot water shines.

We see things now only because visible light waves reach our eyes. But even when objects don't give off visible light, they give off other waves that can be detected in other ways. Objects that are warmer than their surroundings give off heat waves, which are also called infrared waves. Infrared waves are a form of light, except the waves have a longer wavelength than visible light.

Put your hand next to different objects in your house. You can actually feel the heat given off by certain hot objects, such as light bulbs. Some snakes can sense heat in complete darkness and use it to "see" their surroundings. These snakes can tell that animals are nearby because of the heat from the animals' bodies. People have also learned how to use heat or infrared waves to make pictures.

Infrared waves can come in pretty handy. The coast guard has equipped helicopters with infrared detectors that help locate infrared waves from people lost in the sea at night. The army uses infrared guided missiles to detect hot engines of other aircraft. Some homes even have infrared motion detectors that turn on lights when somebody comes near. Although you can't see infrared waves with your eyes, they can be captured on film. You could also use special goggles to make them visible. Soldiers use these goggles to help them see at night.

Into the Wormhole
Time

Someone once said that time is what keeps everything from happening all at once. Time seems like a river that is steadily carrying everything into the future. Once we get to a future tomorrow, today becomes yesterday. But what is time? Can it flow faster or slower? Will it ever be possible to go backward or forward in time? It's fun to wonder about, if you have the time.

ACME
TIME TRAVEL, INC.
"WE GET YOU THERE
YESTERDAY!"

FUTURE
NOW
PAST

What if all past time were squeezed into one year

You probably think of a year as being a very long time. But the universe has been around about 15 billion years, and the Earth itself has been around for almost 5 billion. A billion is such a big number that it's hard for most people to imagine it. (See the box on the next page to get an idea.) By pretending that all past time is squeezed into a single year, we can begin to understand just how long 15 billion years is.

If 15 billion years were squeezed into one year, the year would start off with a real bang—the biggest New Year's party ever. Scientists believe that our universe began in an enormous explosion called the **big bang** that occurred around 15 billion years ago.

In the early universe, there weren't any planets, stars, or galaxies. There were only light and particles whizzing around at very high speeds. Stars and galaxies formed when particles collided to form bigger clumps. These clumps then attracted each other by their gravity.

SPARE **T**IME.   Do you know how long a year is? It's not 365 days. It's 365 days, 5 hours, 48 minutes, and 46 seconds long. That's how long it actually takes the Earth to go around the Sun. The extra 5 hours, 48 minutes, and 46 seconds amount to roughly one-fourth of a day. How do you think the early calendar makers accounted for this spare time? They invented leap years, special years in which February has 29 days instead of the usual 28. Those special years are one day longer and occur every 4 years. ◀

64

ONE **M**ISSISSIPPI, **T**WO **M**ISSISSIPPI.   Even a million—which is 1,000 times smaller than a billion—is a very big number. If you wanted to count to a million, it would take you around two weeks. You'd have to count day and night if you were counting one number every second. You could count to a billion, but don't plan on doing anything else for the next thirty-two years! ◀

If all of time were a year, stars and galaxies formed around March. Our own solar system, including the Earth, the Sun, and the other eight planets, formed around September. And the first life on Earth probably developed around October.

It wasn't until December, the last month of the imaginary year, that the first mammals showed up. Humans didn't appear until the last minutes of the year! Cars, electric lights, and airplanes were just invented about a fifth of a second ago—which is about how long it takes to blink your eye.

JANUARY

FEBRUARY

MARCH

APRIL

MAY

JUNE

JULY

AUGUST

SEPTEMBER

OCTOBER

NOVEMBER

DECEMBER

DECEMBER 31, 11:59 PM

DECEMBER 31, 11:59:59 PM

Not Much Going On Here

What if you could travel into the future

For many people, just seeing a glimpse of the future isn't good enough. They want to go there. Traveling into the future has been the subject of science fiction books and movies for years. Imagine that you could just step into a magic machine, set a few dials showing the year you want to go to, and the machine would take you there. You'd be able to see how your future life turns out, or how the world turns out. Of course, you'd want to be sure to have a round-trip ticket!

Guess what. There may be a way to travel into the future! By using freezers much colder than your home's fridge, scientists have been able to put certain animals into a deep freeze. Time stands still for the frozen crea-ture until it is defrosted and revived. The creatures aren't dead when they are frozen, but their bodies don't age. Of course, time really doesn't stand still in this case. It's as if the creature goes to sleep and wakes up in the future. Someday, people may be able to travel into the future this way. For now, though, scientists have not discovered how to make frozen bodies come back to life.

Another way to travel into the future would be to travel at a speed almost as fast as the speed of light. When things travel close to the speed of light, time slows down. Fast-moving clocks run slowly. Fast-moving people live in slow motion—at least according to someone watching their spaceship go past. But it's not an illusion that time slows down, even if someone in the moving space ship wouldn't notice it. They'd definitely notice it after they returned to Earth.

If you went fast enough, you might find that while you were gone for a week-long trip, a million years went by according to clocks on Earth. Yet the clocks in your spaceship would show that the trip lasted only a week! The Earth would look pretty unfamiliar when you returned, and your friends wouldn't still be around.

Of course, you shouldn't plan on a trip like this just yet. We probably won't have spaceships that can travel near the speed of light for a long time, if ever.

TRANS-TIME AIRWAYS

DEPARTURES

FLIGHT NUMBER	DESTINATION	LEAVING	ARRIVING	GATE
2300	HERE, PLUS 1 HR	1:35 PM	2:35 PM	A
2301	23RD CENTURY	5:15 PM	5:16 PM, 2300 AD	G
312	ANCIENT ROME	11:01 AM	11:02 AM, 30 AD	B
551	1776, AMERICA	2:02 PM	2:03 AM, 1776	E
765	YESTERDAY	3:35 AM	3:36 AM, YESTERDAY	J

ARRIVALS

FLIGHT NUMBER	ORIGIN	ARRIVING	LEFT	GATE
2300	HERE, MINUS 1 HR	2:35 PM	1:35 PM	A
716	MESOZOIC ERA	12:05 PM	65,000,000 BC, 1:10 AM	D
765	TOMORROW	3:36 AM	3:35 AM, TOMORROW	J

CINDY

EMPLOYEE
OF NEXT
MONTH

H. G. WELLS

TRANS-TIME AIRWAYS

WHERE TIME FLIES FIRST CLASS

What if you could travel into the past

Do you ever wish you could travel into the past and do things differently? Or maybe travel back hundreds of years to see what it was like then?

If you really could go back and change the past, you could meet your younger self. You could give yourself all kinds of good advice. But your younger self probably wouldn't believe that you were from the future. Would *you* believe an older person who claimed to be a future version of you?

By going back and changing the past, you might even change the present so that you never existed! Suppose, for example, that you accidentally did something that resulted in your parents not meeting each other? But if they never met, and you were never born, then how could you

go back in time at all? Such problems make most people think that the idea of going back in time is impossible.

▼ _____

B**LACK HOLES.** Suppose gravity on Earth were much stronger than it actually is. It would be as if a giant hand squeezed the Earth into a tiny ball. Very big stars have much more gravity than the Earth has. While a star is shining, it gives off enough heat to keep gravity from squeezing it into a smaller size. But once a big star burns out, its own gravity squeezes it into a tiny ball. It has become a black hole. Its gravity is so strong that nothing—not even light—can escape from a black hole. ◄

_____

There are other reasons why most people think that you can't go back in time. If somebody in the future had learned how to travel in time, wouldn't we have noticed time travelers from the future? Even if they didn't want to be noticed, you'd expect to find some evidence of time travelers' visits. You'd think that many important events in the

"Don't mind me—I'm just a time traveler."

past, such as the signing of the Declaration of Independence, would be big "tourist attractions" for time travelers.

Most people don't think that you could really travel to the past. But some scientists are not so sure. In the universe, when very big stars burn out, they become black holes. Some scientists think that certain black holes can be used for time travel. Such black holes are called **wormholes.**

A wormhole on Earth is a path dug by a worm that connects places in the ground. The wormholes formed from stars are supposed to connect two different times and places in space. If you fell into a wormhole like that and survived, you'd supposedly come out at another place and another time—maybe in the past. Is this really true? Nobody knows for sure if there are such things as wormholes.

Candle Clocks and Supercomputers
Inventions

All the great inventions in our society were made by a person or a group of people who first asked themselves a "what if" kind of question—such as, What if light could be made from electricity? After an inventor gets a great idea for an invention, he or she often has to spend a lot of time turning that idea into a something that works well. A number of people had the idea for making a lightbulb using electricity, but only Thomas Edison (1847–1931) spent enough time on it to make that idea and many others work. Great inventions such as Edison's lightbulb have completely changed society and have also led to a steady stream of new inventions. No one would have thought up the idea for making television or motion pictures if we didn't already have electric lights. Some inventions can even change the way we think about ourselves. After computers were invented, they began to do things that only people used to be able to do. Some people began to wonder if computers could think, and what it means to be able to think. What do you think?

What if cars didn't need fuel?

If cars could run without any fuel, all the gas companies and gas stations would go broke. But you'd never have to fill up the gas tank or worry about running out of gas on a deserted highway.

For now, though, most cars get their energy by burning a fossil fuel, such as gasoline. (See the box on the next page.) Burning the fuel releases the fuel's energy in the form of little explosions inside the car's engine. The little explosions make the engine turn. The engine turns the car's wheels to make it go. As the car burns fuel, however, polluting chemicals come out of the tailpipe at the back.

GASCO

CLOSED

If we want to eliminate pollution from cars, we need to develop cars that run on a different kind of energy. **Solar** cars get their energy directly from the energy in sunlight. Special solar panels on top of the car change the sunlight into electricity, another type of energy, that makes the car go. The sunlight provides energy for free, but solar cars are still very expensive to build and can't go very fast—yet.

Someday, most cars will probably get their energy from electricity. Electric cars do not cause any pollution. Large batteries supply the electricity for these cars. Batteries have chemicals inside them that store energy. When these chemicals react, they produce electricity. as a result, batteries lose energy, or run down. The battery of an electric car needs to be recharged each night by plugging it into a source of electricity.

Of course, electricity isn't free either, and creating it can cause pollution. Electricity comes from an electric power plant that probably burns some fossil fuel in order to produce the electricity. So, electric cars don't create pollution directly, but they can indirectly cause pollution at the electric power plant.

IT KEEPS GOING AND GOING AND GOING AND GOING.   Energy is what makes things go and what gives things the ability to do work. There are different forms of energy, including heat, light, chemical energy, electricity, and energy of motion. Energy can't be created or destroyed, only changed from one form to another. So cars will always need an outside source of energy in order to work, but the type of energy needed may change.

Most forms of energy originally come from the Sun. Living things—both plants and animals—store up energy from sunlight. When they die, plants and animals are buried underground. There heat energy changes the material into fossil fuels—coal, petroleum (oil), and natural gas. It takes millions of years for these fuels to form. Once used, they cannot be replaced. It will take millions of years for more to form. We shouldn't use them up too quickly. ◀

What if electric lights hadn't been invented

Have an adult light a few candles for you one night, and turn out all the electric lights. You'll see how dark it was before electric lights were invented. You would probably be barely able to read a book by candlelight. Of course, for some things, such as watching TV, you don't need a whole lot of light. But if there were no lightbulbs, there certainly wouldn't be any TV!

The lightbulb was invented by Thomas Edison just a little over 100 years ago. Before lightbulbs, people used candles, oil lamps, and gas lights for indoor lighting at night. The candles, oil lamps, and gas lights often tipped over and caused fires.

Edison also figured out how to make electricity at a power plant and deliver it to people's houses. (See Keeping Current.) Once electricity was installed in people's houses, it was used for all kinds of other purposes besides providing light.

How do electric lightbulbs work? The main part of the lightbulb is the **filament**—a very thin piece of wire inside the glass bulb. When electricity flows through the filament, it gets very hot—so hot that the filament begins to glow. You can see the

KEEPING CURRENT. Thomas Edison used electricity to make his lightbulb give off light. Electricity is a form of energy produced by the movement of electrons, the tiny particles inside atoms. The movement (flow) of electrons through a substance, such as a wire, is called an **electric current.** ◄

same thing happen to the wires in your toaster. They glow bright orange when the toaster is turned on. The filament of a lightbulb gets even hotter than the wires in your toaster, which is why the lightbulb glows much more brightly.

Nowadays, we have lightbulbs that don't even have filaments—for example, **fluorescent lights.** Fluorescent bulbs are coated on the inside with a substance that glows when exposed to ultraviolet light. They don't get as hot as lightbulbs with filaments, because more of the electrical energy is changed into light, and less energy is changed into heat.

The newer types of lightbulbs save energy. They produce the same amount of light as filament bulbs but use less electricity. Using fluorescent bulbs is one way to help our environment, because they use less energy.

What if clocks hadn't been invented?

You might think that a world without clocks would just mean that you'd have no annoying alarms and no way to tell if it's time to do your homework or to go to bed. Think again. A world without clocks would also mean no computers (or computer games!), no radio, and no TV.

The technology used to build clocks is also used in many of the devices we take for granted in everyday life. Computers use an internal clock in order to work, and they could never have been invented without clocks. Without the technology that makes clocks possible, we wouldn't have any radio or TV either.

Our world would be very different without clocks. Explorers might never have arrived in America. Traveling across the oceans by ship would have been almost impossible without some means of telling time. Early sailors used the positions of the stars to find their location at sea. But the stars move across the sky at night as the Earth rotates. In order to use the stars to find your location, you need to know what time it is.

Thousands of years ago, people learned to use the Sun to keep track of time. They used sundials, which were really just straight sticks stuck in the ground. As the Sun sweeps across the sky, the stick's shadow moves. People used the position of the stick's shadow to keep track of the time of day. Sundials aren't much good on cloudy days or at night, so people used them to mark off the hours on candle clocks. When the candle melted from one mark down to the next, they knew that an hour had passed.

The earliest mechanical clocks were invented about 600 years ago. But they couldn't tell time very well. Clocks improved about 300 years later when the Italian scientist Galileo Galilei (1564–1642) discovered a secret about swinging pendulums. Galileo discovered that each swing of a pendulum lasts exactly the same length of time. So to keep track of time, all you need to do is count the number of swings of a pendulum. Better yet, if you could make a clock with a pendulum inside it, the clock could count the pendulum swings by itself. The clock moves its hands a little bit every time the pendulum swings, and the hands show the time. (You can see swinging pendulums today in grandfather clocks.)

As Time Drips By *Making a candle clock.* Get two candles that are long and thick. Put the candles in holders next to each other. Have an adult light one of the candles. When one hour has gone by the lit candle will be shorter. Make a mark on the wax of the unlit candle at the place where the top of the lit candle comes up to now. Repeat each hour until the lit candle is all gone. Your unlit candle is now a candle clock. You should be able to tell how many hours have passed by the marks you've made in the wax. Have an adult light your candle clock. Watch time drip by to see how accurate your clock is. You're probably glad that you have a watch!

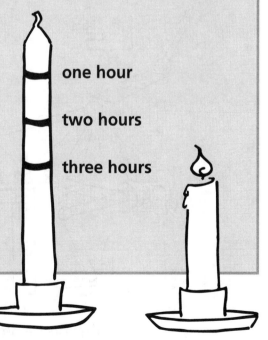

one hour

two hours

three hours

What if computers could think like people?

In movies, supercomputers sometimes take over the world. For now, in the real world, computers can only do what people tell them to do.

The people who tell computers what to do are called **programmers.** They write the instructions (the program) that the computers have to follow. Computers can function like a brain in robots, directing the robot's actions. Today, there are some elaborate computer-controlled robots that perform very complicated tasks. Robot vehicles called land rovers collect data from the surface of Mars and radio it back to Earth. Other types of robots work in assembly lines to build automobiles.

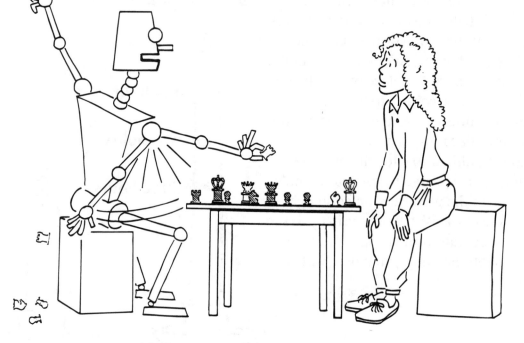

"Observe! Isn't that a comet?"

THE TURING TEST. Alan Turing was a famous computer scientist who designed a test to see if computers seemed to be thinking. Suppose a computer was programmed to respond to your questions about anything you wanted to ask. Its answers would try to fool you into thinking you were talking with a real live person, not a computer. In the Turing Test, people typed questions into a computer and saw the answers on the screen. They had to guess if they were communicating with a real person at a hidden keyboard or a computer in the next room. If it was a computer, and it did a really great job, they might think they were communicating with a person. Suppose you guessed wrong as often as you guessed right. Then, according to Turing, the computer could be considered to be thinking. But some people have said that the Turing Test doesn't really tell whether computers can think. It only tests whether computers can make people think they are thinking. Other people reply that the test is just fine, because that's the only way we can tell that other *people* are actually thinking. What do you think? ◀

There's even a robot land rover that uses TV camera eyes, lasers, and radar. It can change its course to avoid an obstacle in its path. In a sense, the robot can learn about the world in much the same way that people do. Its computer "brain" changes depending on new data from its environment. The robot then adapts its behavior based on what it has learned about the world. But is the computer brain really thinking? No. People have to tell the computer in the rover how to make choices about whether to go left, right, or straight ahead.

Scientists disagree about whether a computer could ever think. It comes down to the definition of "thinking." Computers can now do some things much better than most people can— for example, doing arithmetic, tracking a space satellite, or playing the game of chess.

But people do many things much better than a computer can, such as recognizing faces or having a conversation. To find out if computers do think, scientists have put them through thinking tests. Read about the Turing Test in the box on this page.

Smart Pigs and Pet Dinosaurs

Plants and Animals

Life on Earth has evolved, or developed, because organisms (plants and animals) try to take better advantage of their natural environment. For example, animals in a given environment often take on colors that allow them to blend into their environment. This is called **camouflaging.** This way they can avoid becoming a meal for other animals. But these changes don't happen because of any planning. An animal doesn't say "Gee, maybe if I were a different color, I'd blend in better."

Instead, each time animals have babies, some of the babies have characteristics that allow them to survive better than others. The ones that survive better pass those characteristics along in their **genes** to the next generation when they have babies. In this way, organisms gradually evolve to become better adapted to their environment. Sometimes big changes can occur to organisms over a long period of time, and a new type of organism (a species) is created. Organisms can adapt to their environment even if it is changing slowly over time. But if the environment changes too rapidly, a species of plant or animal may find itself in big trouble. The dinosaurs, for example, were very successful animals during the millions of years they lived on Earth. But they were all wiped out— probably as the result of a dramatic change in the Earth's climate that happened 65 million years ago. Most people think humans are smarter than other animals. If we really are smarter, maybe we can find some way to keep the Earth's environment a pleasant place to live.

What if the dinosaurs hadn't died out

We might have road signs that say CAUTION: DINOSAURS CROSSING. Triceratops stew could be a favorite family meal. Or maybe we wouldn't even be here. If the dinosaurs hadn't died out, early humans might have had a tough time surviving.

Every one of the dinosaurs has disappeared from Earth, so we say they are **extinct.** Some people think that there must have been something wrong with the dinosaurs that caused them to become extinct. But the dinosaurs lived successfully on Earth for a much longer time than humans have so far.

So why did the dinosaurs die out before humans appeared on the scene? No one knows for sure. The best guess is that they died out because the Earth's climate underwent a big change. Many scientists think this big climate change happened about 65 million years ago when a big meteorite hit the Earth. A very large meteorite would have created a big cloud of dust in the atmosphere, which would have cooled the planet. The cooler climate could have killed the dinosaurs and most other large animals around at the time.

How come big animals like the dinosaurs had a tough time surviving? In nature, there are more small animals than big ones. Big animals eat smaller animals or lots of plants. The bigger the animal, the fewer there can be in a given area, because there just isn't enough food to go around.

Even if the dinosaur didn't die out, humans might still have survived. Our intelligence could

have made up for the greater strength and size of the dinosaurs. Early humans successfully hunted some very big animals, which are now extinct.

Even if dinosaurs hadn't died out 65 million years ago, they might have become extinct after human hunters spread over the planet.

What if we brought the dinosaurs back

You might see signs saying "Baby Dinosaurs for Adoption." But most people probably wouldn't want to adopt baby dinosaurs. Not only would these babies need a really big backyard, but some might also grow big enough to destroy their owners.

In the movie *Jurassic Park*, scientists found some dinosaur DNA in a mosquito that had been preserved in amber, and they used it to create living dinosaurs. **DNA** is a molecule found in every living cell. It contains the code that tells how any living thing is built. (See Creation's Code for more information.) The scientists put the dinosaur DNA in ostrich eggs in order to hatch live dinosaurs. When the eggs hatched, the sci-

entists found that the dinosaur DNA had made baby dinosaurs.

You might say "Well, that's just a movie!" Scientists can't make dinosaurs like stegosauruses and pterodactyls

CREATION'S CODE: DNA. DNA is a molecule found in all forms of life, including dinosaurs and people. Your DNA gives the basic plan for many things about you—your eye color and hair color and more. As you grow up, your body changes based on the instructions it gets from DNA. DNA controls your development the same way that plans for a house control how the final house will look. ◀

today, but they can't say for sure that it's impossible. Even though the dinosaurs have been gone for around 65 million years, some of their DNA might still be around in **fossils** (parts of organisms that have been preserved in the Earth's crust).

Let's say we did bring the dinosaurs back. How would they survive without momma and poppa dinosaurs? And how would we know what to feed them? Some scientists have tried to figure out what dinosaurs ate by looking at fossils of dinosaur waste, but many of the foods that dinosaurs ate don't even exist anymore!

Even if we knew what to feed baby dinosaurs, it would be difficult to raise them in today's world. Very large animals like the dinosaurs need large areas of open space—forest, jungle, or grassland—in which to roam around. They probably wouldn't be happy being kept in a zoo just so people could watch them. As in the *Jurassic Park* movie, dinosaurs could also cause a lot of problems if they escaped.

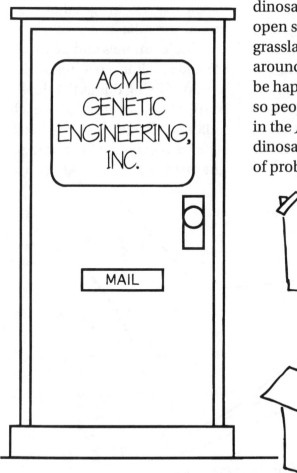

ACME GENETIC ENGINEERING, INC.

MAIL

FREE BABY DINOS ↓

What if all plants and animals disappeared

S quish. Swat. Stomp. You probably think nothing of killing a cockroach, swatting a fly, or stepping on ants. But what if all insects were no longer here one day? Or all animals and plants? It's pretty certain that you wouldn't be here either.

Humans rely on the animal and plant world much more than we realize. Without plants and animals, we'd have nothing to eat. Without plants, we couldn't breathe. During the process called **photosynthesis,** plants take in carbon dioxide (the gas people breathe out) and create oxygen (the gas that people breathe in).

Could animals and plants disappear almost overnight? A group of living things that look similar and can breed with each other is called a species. When a species is extinct, it means that they've all died out, that not a

"Well, there goes my last chance for a date on Saturday night for the rest of my life."

single one is left. All throughout the Earth's history, species have become extinct. Sometimes a lot of species become extinct all at the same time. This is called a **mass extinction.** Sixty-five million years ago, there was a mass extinction when the dinosaurs and many other species disappeared at around the same time.

These mass deaths probably happened because of major changes in climate. But today one of the greatest causes of extinction for plants and animals is humans. When people do such things as clear land for farms or buildings or highways, the plants and animals living on that land often die. Some die quickly, and some die slowly. The ones that die slowly can't adapt to the new environment fast enough. The most dramatic example of this is found in tropical rain forests. (See the Incredible Shrinking Rain Forest.)

THE INCREDIBLE SHRINKING RAIN FOREST.   Near the equator, halfway between the North Pole and the South Pole, there's a hot, wet region called the tropics. Rain forests are tropical woods that occur in a band around the Earth, sort of like a belt. Although rain forests take up only 6 percent of the Earth's surface, they contain more than half of the Earth's species. Just a few acres of the Amazonian rain forest in South America hold more kinds of plants than are found in all of Europe.

But as humans clear the land for farms and pasture, the rain forest is disappearing. Every 60 seconds, about 100 football fields worth of rain forest burn. At least 4,000 tropical species become extinct there each year. These animals and plants may be far away from us, but they still affect us in many ways. The plants of the rain forests take in carbon dioxide from the air and produce oxygen. Doctors get valuable lifesaving medicine from rain forest plants.

On the other hand, the people of the tropics need to make room for new roads, farms, and houses. The solution to the conflict between the needs of the people of that area and the desire to protect the environment will not be easy. ◀

The Amazing, Changing Me
People and Animals

Most scientists believe that millions of years ago there was a creature that evolved to become present-day humans. That same creature also evolved to become present-day apes. So humans and apes have evolved from a common ancestor. During their evolution from that ape-like creature, humans began to walk upright, and they lost most of their hair. But the biggest changes occurred when humans started using tools and learned to change their environment. Humans have been able to do things that no other animal on Earth can do: write books and build a technology that includes cities, airplanes, television, computers, and even spaceships that can visit the Moon. If humans are just animals, how have they been able to do all these amazing things? Most scientists would say that people are way ahead of other animals because of their superior brains. But some animals (such as porpoises, whales, and elephants) also have very big brains. Maybe it's people's *hands,* which can manipulate their environment, that give people their big advantage. Of course, some animals are able to do things that people can't. It would be nice if we could fly like a bird, walk upside down on ceilings like an insect, or maybe be as big and strong as an elephant.

What if some animals were smarter than people

You might be behind bars at the local zoo while some chimpanzees take your picture and throw peanuts at you. Or maybe you'd be your dog's "pet human." Even worse, you might be raised as a tasty meal for lions and tigers, or you might come in handy at the lab when some elephants want to test a new shade of elephant eye liner. In other words, if some animals were smarter than people, they would take our place as the creatures-in-charge on Earth.

You probably wouldn't want this to happen. Thinking about this probably makes you feel bad about how humans treat animals. People, however, are not the only animals that **dominate** (control) other animals. Many animals eat others for food. Ants

What if we're not the smartest animals on Earth?

We think that people can do things that animals can't, such as write books or build buildings. But some insects, birds, and animals, such as beavers, build very elaborate structures. No animals have developed the kind of complex technology that people have, but are people today really any smarter than people were 10,000 years ago when they had no technology? And who knows, maybe some animals have a complex language that we have no idea how to read. Maybe whales tell each other long stories in whale-talk. Most people think that humans are the smartest species, but they can't prove it. Think about it—any test we make up to see how smart different animals are will certainly show us to be the smartest, because *we* made up the test.

even keep bugs called aphids for food. Some species, such as chimpanzees, even sometimes have pets.

Scientists believe that some animals—for example, chimpanzees, elephants, whales, and porpoises—are very smart. Some chimpanzees have learned over 200 words and can carry on a conversation using signs instead of speaking. Chimpanzees even use simple tools to get food. Scientists used to think that only people used tools.

Whales sing complicated songs that can sometimes last for an hour. Porpoises also seem to have a very complex "language." But scientists really have very little idea what porpoises or other animals say to each other. Some animals might actually be very smart—in a different way than we are—and we don't know it because we can't communicate with them.

What if you were the size of an insect

You could walk on the ceiling on your tiny sticky feet. You could jump 20 times your own length in one bound. You could carry some-thing 300 times heavier than yourself. Of course, you could also get stomped on by a giant foot as you walked across the sidewalk!

If you were the size of an insect, you could do amazing things. For example, a flea leaps 12 inches (about 30 cm), which is 200 times the length of its own body. That would be like a person jumping the length of five city blocks. A bee can haul a load 300 times its own weight. That

WHY SMALL CREATURES ARE STRONGER FOR THEIR WEIGHT. Think of the muscles in a creature's body as being like a stretched rubber band holding a weight. Compare two rubber bands, one of which is twice as long and twice as wide as the other. If we keep all proportions the same, the weight hanging on the end of the longer rubber band is twice as long, twice as wide, and twice as deep as the other one—so it's $2 \times 2 \times 2 = 8$ times as heavy. Suppose the weight on the thin rubber band is the most that it can support. Then the 8 times heavier weight on the thicker rubber band would surely break it.

If you could be reduced to the size of an insect, keeping your proportions the same, you would be as strong as it is. You could do the same kinds of things that insects can do—like walk on the ceiling. How come? A tiny version of you has much less muscle power than the regular-size version. But, for its weight, the tiny version is much stronger. Its sticky feet allow it to walk upside down on the ceiling, without falling. Don't get any ideas. You can put all the glue you want on the bottom of your shoes, but a normal-size you couldn't walk on the ceiling. You're just not strong enough for your weight. ◀

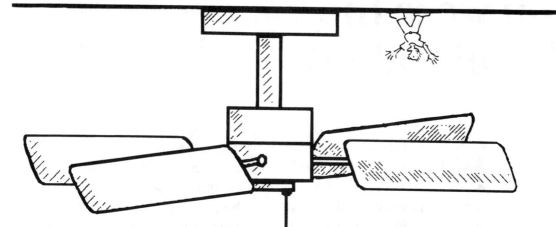

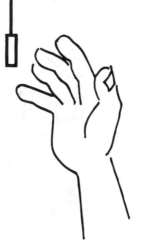

would be like a human pulling three heavy trucks at the same time. How do insects do these things?

The key is to understand how an animal's size, weight, and strength are related. Compare two types of animals whose sizes are very different—say a man and a tropical stick insect. A man is ten times as long as a tropical stick insect, and much stronger. But a man weighs hundreds of thousands times as much as the insect. For any two creatures having different size, the smaller one has more muscle power for its weight. So, insects are much stronger than people for their weight.

What if you were the size of a giant?

Remember fairy tales like "Jack and the Bean Stalk"? In these stories, the giants are powerful and frightening and can do amazing things. Yet if you were actually a giant, you might be amazed at all the things you *couldn't* do. Just getting out of bed in the morning would be a major effort.

If you were a giant 10 times your present size, you'd be able to take much bigger steps. But because of your heavy weight, you might not be able to run much faster than you can now—or even at all. You couldn't stand up quickly or jump because your heavy weight would pull you back to Earth. You'd be as clumsy as an elephant. Once you got up, you'd have to be careful not to fall down, because the heavier and larger you are, the harder you fall. You'd be more likely to get hurt because of your heavy weight.

BIG BLUE.   The largest animal in the world today doesn't have to stand up. One blue whale weighs as much as about 24 big elephants. A creature that floats in the water doesn't have to support its weight like land creatures do. Also, getting enough food to eat is less of a problem. Every day for a blue whale is an all-you-can-eat seafood feast. The blue whale swims with its mouth open to take in tons of water and tiny shrimp. ◀

The pull of gravity on a body is called its **weight.** Just as gravity affects an insect much less than it does a person, it would affect a giant much more. Have you ever noticed that very large animals such as elephants have thicker legs for their size than smaller animals have? Very big animals need those thicker legs to support their weight. Suppose you were enlarged to 10 times your present size, but had the same proportions. You would be 10 times taller, 10 times thicker, and 10 times wider, and you would weigh 10 × 10 × 10 or 1,000 times as much. You would collapse under your own weight, even with your 10 times thicker legs. That's because your legs would need to support your weight, which is 1,000 times more than it is now.

And what would your 10-times-bigger self do all day? Getting enough to eat would take so much time that you couldn't do anything else. The bigger an animal, the more it needs to eat to stay active. At 10 feet (3 m) high, elephants are the largest land animals still around. Do you think that they spend time chatting on cellular phones, playing sports, or going to school? Nope. They spend almost all day eating. Big dinosaurs were somewhat bigger than elephants. But no animal could grow to weigh 100 times more than an elephant. They couldn't possibly get enough to eat, and they probably couldn't even stand.

What if you had a dog's sense of smell

Boy, it sure would be one smelly world! If you had a dog's nose, you'd smell other people's sweat, even though they walked past you hours ago. You'd smell trash buried underground. And your nose would probably be bigger, wetter, and noisier!

We don't rely on our sense of smell as much as we do our senses of sight and hearing. But, if your sense of smell were as good as a dog's, you'd probably use it much more. Dogs and other animals use smell to alert them of danger and to find mates or food. Dogs even mark the boundaries of their territory with their scent, so that other dogs can use their sense of smell to tell whether they should stay away.

WHAT PERFUME IS THAT? *The creature that has the very best sense of smell is the silkworm. A male silkworm moth can tell if a female is around even if she is 7 miles (11 km) away. But, unlike dogs, silkworms can only smell their mates—nothing else.*

IT'S A SMELLY WORLD. *A dog's sense of smell is a million times better than yours. Suppose a dog could smell something if five molecules reach its nose. Then you couldn't smell it until 5 million molecules reach your nose.*

You smell something only when molecules in the air reach an area inside your nose that sends a signal to your brain. For both dogs and people, different molecules fit into different places in the nose, like different keys fit different locks. The kind of smell that you sense depends on the particular place where a molecule fits in your nose.

"Great! Mom's baking chocolate chip cookies!"

Most smells are a combination of many different kinds of molecules. For example, the smell of underarm sweat involves 250 different kinds. A dog can recognize the smell based on the particular combination of molecules. It's sort of like the way your brain recognizes a piece of music or someone's voice. Your brain recognizes the particular combination of notes.

THE NOSE KNOWS. Before police use a bloodhound dog to track criminals, they test the dog to see how well it can smell. Here's how they tested one bloodhound: A police detective walked across New York City's Central Park. The night before, 55,000 people had been in the park at a rock concert. Then the policeman went across a meadow where nine baseball games were being played. Finally, the policeman walked up a street where many people were walking dogs. Then, the trained bloodhound was given the policeman's jacket to sniff. Following the invisible trail of molecules, the dog took only five minutes to find the policeman! ◄

What if you could fly like a bird?

Wouldn't it be wonderful if you could flap your wings and fly anywhere you wanted? You wouldn't need to worry about following any roads or getting stuck in traffic. But the sky might get pretty crowded if everyone could fly.

For thousands of years people have looked at birds and dreamed of flying. People have even strapped on large wings to try flying. Thump! They always came right back down again. Only in the past 100 years have we been flying in airplanes. Now we have superlight airplanes without any engine that a single person can fly just by their own power. A person flying this special plane pedals a kind of bicycle that turns the plane's propeller. But you need to be very strong to keep the propeller going fast enough to fly the plane very far.

There are three good reasons why you can't just strap on a pair of wings and fly. First, most people's arms are not very strong. Second, we haven't figured out how to make a pair of wings that

How Do Airplanes Fly?

How Do Airplanes Fly? Airplanes fly partly based on something called Bernoulli's Principle, which says that an air flow creates a low pressure. The shape of an airplane's wing makes the air move faster over the top than over the bottom, so the pressure is lower on the top. The higher pressure on the bottom of the wing pushes the plane upward. To see Bernoulli's Principle in action, get two balloons, and hang them by strings, so that there's a small space between them. Blow hard into that space, and the two balloons will come towards each other, because the pressure is lower where the air flows fastest.

98

move quite like a bird's. Airplane wings don't flap up and down. They use a propeller or jets to move forward through the air.

The third and most important reason you can't just strap on a pair of wings and fly is that you're just not strong enough for your weight. Most birds are very light and their bones are hollow, so their flapping wings are able to keep their bodies in the air. But the bigger a bird gets, the longer its wings need to be in order to fly. It takes more strength to flap very big wings than it does to flap small ones. That's why there are no flying birds as heavy as people.

FORMER FLIERS. *Some big birds such as penguins and ostriches have lost the ability to fly. Penguin wings have evolved to become flippers that help them fly through the water. Ostriches, the largest living birds, can run at great speeds.*

What if you saw the world upside down

The floor would be above you. Mountains would point down. You might even have a hard time recognizing people you know if you saw them upside down. (Check it out—try turning some photos of your family upside down.) Feel dizzy yet?

But here's something really strange. Your eye actually sees the world upside down all the time! There's a lens at the front of your eye. Light enters this lens and forms an upside-down image on a "screen" at the back of your eye. This screen is called the **retina.** The light reaching your retina is changed to a message that goes to your brain. Your brain flips the image on your retina, and you see right-side up.

There's something else that makes upside-down pictures of the world. It's what you used to take those family pictures. The camera is designed on the same principle as the human eye. In a camera, light from outside objects enters the lens. It then produces an upside-down image on the film at the back of the camera.

Your brain's ability to make sense of the world is truly amazing. Scientists have even done

Magnifying Magic *How an upside-down image is formed.* Get a magnifying glass and a sheet of paper. Tape the paper onto something stiff, such as a book or a piece of cardboard. With your back to a lamp, hold the paper so that light from the lamp can reach it. Hold the magnifying glass in front of the paper. You should see an upside-down image of the lightbulb on the paper. If you don't see the image, try moving the magnifying glass closer or farther from the paper until you see it. You may also be able to see an upside-down image of yourself if you look at your reflection in the hollow side of a spoon.

experiments where they put special glasses on people. These glasses make the world look upside down. Let's say you put on a pair of these glasses. Everything would suddenly look very strange. You'd probably bump into things while walking around. But, here's the weird thing: after wearing these glasses for a few days, you'd begin to see the world rightside up! Your brain knows how to make an upside-down world look normal, so you don't even notice it. Scientists don't know how the brain does it.

What if you had three eyes?

Guess what—you do! We all have a kind of third eye in the middle of our brain. This "third eye" is called the **pineal gland.** It doesn't really see anything, but it senses light. Its main purpose is to control our moods. It also keeps the body's processes in step with the 24-hour daily cycle. Some crea-tures, such as certain fish, frogs, and lizards, actually do have a real third eye that allows them to see in a different direction from the other two.

Where would be the best place to put a real third eye if you had one? How about in back of your head? If you had a third eye there, you could see in all directions at once. You could see if someone was sneaking up on you without turning your head. Or how about putting your third eye on the end of your finger? Then you could put your finger next to something and get a closer look.

If you put your finger "eye" at

Thumbs Up *How we judge distance.* Hold your thumb at arm's length. Close your left eye and look at your thumb and another object far in the distance. Now, keep looking at your thumb, and at the same time open your left eye and close your right. Your thumb appears to move compared to the distant object. Your thumb seems to move because your eyes see your thumb from slightly different directions. Now repeat the experiment with your thumb closer to your face than before. It will seem to move farther this time when you look at it with different eyes. That's because your two eyes point in very different directions to see your thumb when it's closer to you. Your brain uses the difference in direction each eye looks in to figure out how far away an object is.

arm's length, that eye would be very far away from either eye on your head. That large spacing between eyes would give you a much better ability to judge how far away things are. How come? It's hard to judge the distance of objects that are very far away. That's because your eyes are pointing almost in the same direction when you look at a distant object. But if your eyes were much farther apart, the directions for each eye wouldn't be so similar, and you could judge distances more easily. (Check out the Thumbs Up experiment to see how we judge distance.)

AaBbCcDd

$$2x + 3y = 10$$

$$y = 2$$

$$x = ?$$

What if you looked exactly like everyone else?

How would we keep track of who was who? Maybe everyone would have to wear name tags at all times. Or perhaps we'd develop other ways to tell people apart, such as by smell. But if people wanted to fool you and pretend that they were someone else, they could probably get away with it if everyone looked the same.

Could it really happen that everyone looked the same? Scientists have discovered how to make baby animals that are exact copies using a process known as **cloning.** In normal reproduction, an egg **cell** (the smallest living thing that can exist on its own) from the mother is fertilized by a sperm cell from the father.

ACME SHOES

ONE SIZE FITS ALL!

24

JONES SUITS

ONE SIZE FITS ALL!

27

ALL MED. SHIRT CO

ONE SIZE FITS ALL

SOCK WORLD

ONE SIZE FITS ALL!

WELCOME TO CLONE CITY

What if parents could choose how their babies would look?

The idea of choosing how a baby will look isn't so far-fetched. Scientists already know how to take out and replace certain animal and human genes. The genes are the parts of your cells that contain instructions for the things that make you you—for example, whether you're a girl or a boy, or whether you have red hair or brown eyes.

Some diseases are programmed in a plant's or animal's genes. They get passed on from generation to generation. Certain scientists do genetic engineering. **Genetic engineering** occurs whenever animals or plants are bred to favor certain good characteristics or to avoid bad ones. Farmers do this with crops and livestock. Someday, it might be possible for scientists to fix diseased genes with healthy ones in babies. People may even use this process to prevent bad vision, flat feet, or crooked teeth someday. But what if someone wants to develop a superbaby with superintelligence or supermuscles? Should they be able to do it? What do you think?

The fertilized egg cell has genes from both parents. The cell splits many times to make new cells that eventually become a baby. The baby's genes are a mixture from both parents, so they aren't identical to either parent. In cloning, animals are made from a single cell—not one from each parent—and are thus identical. Identical twins also form from a single cell—so they are clones in a way.

Recently, a sheep named Dolly was cloned. This was the first time such a complex animal had been successfully cloned. Scientists think that humans could also be cloned now. But most people don't think that's such a good idea. Who would decide who got cloned? Would clones be treated as equals, or would they be discriminated against? Is it right to change nature's plan for human reproduction? These are just a few of the moral issues that people are discussing.

What if you always knew what other people were thinking?

You might desperately wish you could turn your mind off! Maybe you'd find out that when your best friend said "That outfit looks nice," she really thought "Wow, that's a funny-looking dress!" Or if everyone else could read minds, none of your secrets would be safe. The person you have a secret crush on would suddenly know and tell everyone!

Some people say that they *can* tell what other people are thinking. The ability to read minds is called **telepathy.** Although certain people claim that telepathy is real, many others are not so sure. Scientists

---

Test Your Psychic IQ *How to find out if you can read minds.* Get a deck of cards, and sit across the table from a friend. Give your friend the deck and have him or her look at one card after another. While your friend is looking at each card, try to guess whether it is a heart, diamond, club, or spade. Have your friend tell you whether your guess is right or wrong. Keep track of how many times you guess right and how many times you guess wrong after going through the whole deck.

There are four possible card suits. So just on the basis of chance, you should guess right 1 out of 4 times. There are 52 cards in the whole deck. So you should guess right around 13 times. On a lucky day you might guess right even more often—maybe 16 times—just by chance. But if you guess right much more often—say 25 times—maybe you can read minds. Actually, you might guess right every single time, even without having telepathy. Maybe your friend just wanted to make you feel special and told you that you guessed every card right even though you didn't. How could you change the experiment so this couldn't happen?

have done experiments to test telepathy. For example, they have one person look at cards, and a second person guess the cards that the first person sees. (Try your own Test Your Psychic IQ experiment.)

None of the experiments so far have shown for sure that telepathy really happens. Often the person doing the experiment can't be certain that the other person isn't cheating. Sometimes the person guessing the cards guesses right more often than usual just because he or she has a lucky day.

Scientists haven't proven that mind reading is possible, but they can't prove that it's impossible. Maybe it's true and happens only very rarely. Some people think that it happens only on very special occasions or with very special people, such as identical twins. What do you think? Oh, never mind, I already know.

Our Sun looks pretty much the same year after year. But over a *very* long time, our Sun, like all stars, will change because of the light energy it constantly gives out. Anything that gives out energy has to change in some way. Like a campfire, the Sun and all the stars must eventually burn out. But, there's no need to worry. It will take *billions* of years for this to happen. By then we'll probably have moved to planets around other stars.

As the Sun approaches the end of its life, it will become what scientists call a **red giant.** It will turn red in color and grow much larger in size. In fact, it will eventually get so big that it will almost swallow up the Earth as it expands. Just imagine what it would be like to look up in the sky and see a red Sun almost as big as the whole sky! Of course, when this eventually happens, the Earth would become very hot. A much bigger Sun gives off a lot more heat than a small one does, even though the bigger red Sun may actually be cooler. Does that sound strange? Just imagine a big pot of hot water on the stove, and the flame from a single match. The match is hotter than the water, but the water gives off much more heat because there's a lot more of it.

What if the Sun doesn't rise tomorrow

It would stay dark as night, roosters wouldn't know when to crow, and people might cry "The world is ending! The world is ending!"

In ancient times, people used to think that **solar eclipses** (when the Sun is blocked by the Moon) meant that the world was coming to an end. They thought this because an eclipse came totally without warning. They thought it was some kind of sign from the gods. Today, we understand how eclipses can occur, and we can predict them.

Actually, the Sun doesn't ever rise—it just looks like it does. The reason that the Sun seems to rise each day is that the Earth

makes one complete turn around its axis each day. As it spins, we see more and more of the Sun until we see it all when the Sun is above the **horizon**—the place where the sky meets the ground. The only reason that we wouldn't see the Sun "rise" would be if something slowed down or stopped the Earth from spinning.

If a very large asteroid struck the Earth in just the right way, it could slow the Earth's spin down to almost a halt. (Large asteroids are like small planets. There are thousands of them orbiting the Sun in our solar system.) An asteroid might someday hit the Earth. But there's practically no chance of a really big one hitting the Earth and stopping the Earth's spin. There aren't any very large asteroids that have the right orbit to hit the Earth and change its spin.

SUN **B**LOCK.   During a solar eclipse, the Sun looks like a flat black disc with a shining halo around its edges. A solar eclipse happens when the Moon passes between the Sun and the Earth, casting a shadow on the Earth. During an eclipse, the Moon's shadow moves across the Earth along a certain path. Anyone in that path sees a total eclipse—where the Sun is totally blocked—as the shadow moves past. ◀

What if the Earth didn't go around the Sun in a circle

Guess what? It's a trick question. The Earth *doesn't* go around the Sun in a circle. Its orbit is really a shape called an **ellipse.** (An ellipse looks like a somewhat flattened circle.) As a result of its "squashed-circle" orbit, the distance from the Earth to the Sun changes during the year—sometimes the Sun is closer and sometimes it's farther away. Many people believe that the changing distance to the Sun is what causes the seasons, but this isn't true. The Earth's tilt affects its climate much more than the changing distance to the Sun during the year affects it. That's why it can be winter in the Northern Hemisphere at the same time that it's summer in the Southern Hemisphere.

Second surprise: *all* of the nine planets in our solar system move in ellipses around the Sun. In most cases, like the Earth's, the ellipses look like circles that are only very slightly squashed. The reason that the planets' orbits are ellipses has to do with the force of gravity. The force of gravity between the Sun and a planet pulls the planet toward the Sun and keeps it in orbit.

To get the idea, imagine a rock tied onto the end of a long string that is twirled in a circle. If the string were to break, the rock would fly off in a straight line because the string no longer pulls on it. In the case of the Earth orbiting the Sun, the force of gravity between the Earth and the Sun acts something like the string keeping the rock in its circular orbit. Without the pull of the Sun's gravity, the Earth would fly off in a straight line and not go in a circle. The Earth's orbit is not exactly a circle, however, because gravity is not exactly like a string between the bodies. It's more like a string that can stretch very easily.

In fact, the pull of the Sun's gravity on a planet gets weaker the farther the planet is from the Sun. If a planet were twice as far

SQUASHED CIRCLES FROM THE PLANET MARS? Hundreds of years ago, people thought that the orbits of all the planets were based on perfect circles. This seemed only natural because the planets are in the heavens, and heaven must be perfect. Johannes Kepler (1571–1630) was a Danish scientist who first realized that the orbits were ellipses rather than circles. He realized this by carefully tracking the motion of the planet Mars. Kepler found that Mars moved in an ellipse, not in a circle. You can draw an ellipse using a piece of string whose two ends are attached to thumb tacks on a piece of paper. The string should be about twice as long as the distance between the tacks. Hold a pen against the string until the string stretches tight, then move the pen around the paper so that the string is always kept stretched. ◀

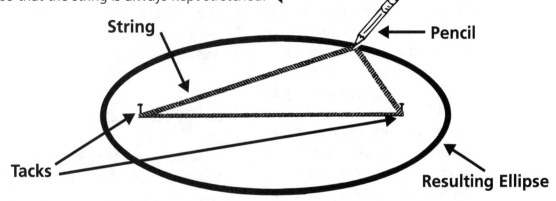

String

Pencil

Tacks

Resulting Ellipse

away, the pull of the Sun's gravity would be only one-fourth as strong. With this kind of a "stretchy string" attached to a rock, it will travel either in a circle or in an ellipse, depending on how you twirl it. It's just the same with the Earth and the other planets and the stretchy string of gravity that connects them to the Sun. Scientists think that the planets were originally formed out of a cloud of dust and debris swirling around the Sun. When the planets were first formed, their orbits could have been either circular or elliptical. But because ellipses can be squashed by any amount, there are *many* more possible elliptical orbits than there are circular ones. So there's practically no chance of a planet's orbit starting out circular when it was formed. All of the planets have elliptical orbits about the Sun.

What if the Earth had two suns?

The days and nights and seasons might be totally out of whack. Or maybe it wouldn't be very different at all. It would depend on how far the double suns were from each other and from the Earth.

The stars in the night sky are actually all suns. Many of them are actually *pairs* of stars close together. They are so close together that they seem to be one star because they're so far away from us. These pairs of stars orbit around each other. If they didn't orbit each other, the pull of gravity between them would make them crash together—and only one big sun would be left.

What if our Sun was one of these "double stars"? What kind of orbit would the Earth have? Here are two possibilities: One is that the Earth would have a much larger orbit than the two suns have around each other.

The two suns would appear close to one another in the sky as they revolved around each other. One sun might pass directly in front and then behind the other as they revolved. This would change the amount and color of the light reaching the Earth.

For example, suppose that one was a big red sun (called a red giant), and that the other was a small white one (called a

ONE SUN, TWO SUN, RED SUN, BLUE SUN. Some **astronomers** (scientists who study outer space) think that our solar system has two suns—not one—and that the two suns are in orbit about each other. Where is the sun we don't see? It may be too faint and far away from our main sun, so we can't pick it out from the background stars. No one knows for sure if this second sun actually exists. But it even has a name in case it's found one day. Our Sun's hidden companion star is called **Nemesis.** ◄

white dwarf). When the white dwarf passed in front of the red giant and then in back of it, the light reaching Earth would change from white to red. We'd need to invent a third word to go with day and night. Maybe we'd say it was "red day" when the bright white sun was behind the big red one, and "white day" when the bright one was in front of it. White day would be warmer and brighter than red day, when everything outside would appear as different shades of red.

Here's the other way that the two suns might appear: The two suns could be far apart and the Earth could be much closer to one of them. In this case, during daytime you'd see the nearer sun in the sky. The farther sun might appear as just a star in the sky at night if it were far enough away. (Some scientists think this might actually be the case—read about it in One Sun, Two Sun, Red Sun, Blue Sun.)

YET ANOTHER HAPPY ENDING ENDANGERED BY THE CHOICE OF WHICH SUNSET TO RIDE OFF INTO.

What if the Moon fell down

Why doesn't the Moon come flying through the sky and crash into Earth like a meteorite? After all, the Earth's gravity is pulling it down, and there really isn't anything holding it up.

Well, the Moon isn't just sitting up in the sky above the Earth—it's moving in orbit around the Earth. The Moon will never fall down, because its orbit around the Earth doesn't change. Isaac Newton was the first to explain why moons in orbit don't fall down.

I Got the Moon on a String *How orbital motion keeps things from falling.* Use a nail to make two small holes in a paper cup on opposite sides near the top. Tie a string through the holes. Put a small amount of water in the cup. Hold the end of the string in your hand while you whirl your arm like a windmill, so that the cup at the end of the string goes in a circle. If you whirl it fast enough, the water won't fall out of the cup, even when its upside down at the top of the circle. It doesn't have a chance to fall down because of the sideways motion of the cup.

Here's how Newton explained it. Imagine that we built a really tall tower—so tall that the top was above the Earth's atmosphere. If you dropped a rock from the top of this tower, it would land near the base of the tower. Now, suppose you threw the rock sideways. It would land some distance away from the base. If you threw it faster, it would land farther away. If you threw it faster than anyone could actually throw it, the rock would miss the Earth entirely and be in orbit around the Earth.

S**ATELLITE STATS.** A **satellite** is any body that orbits a planet. All moons are natural satellites. A man-made satellite is a spacecraft placed in orbit around a planet. Artificial satellites do all sorts of tasks—from collecting weather data to relaying phone conversations and TV broadcasts around the world. *Sputnik 1* was the first artificial satellite ever put in space by humans. It was launched by the former Soviet Union in 1957.

The Moon is a natural satellite of the Earth, so Newton's idea also explained one way to put a man-made satellite in orbit. If you just give a satellite enough sideways motion, it will never hit the ground. It's the same with the Moon. The Moon is always falling *around* the Earth, so it can't fall straight down.

We don't put satellites into orbit from the tops of tall towers. Instead, we launch them on rockets heading straight up from the surface of the Earth. As they rise in the sky, the rockets turn and go sideways. It's this sideways motion that makes the satellites keep falling around the Earth, not straight down. ◀

What if the Moon were much closer to the Earth

If the Moon were closer to the Earth, it would look much bigger in the sky, and its features would become very familiar to us. Nobody would ever say "Can you see the Moon?" or "Where's the Moon tonight?" They'd say "Oh, look, there's the Copernicus Crater!" or "There's the Sea of Tranquillity."

How big an object looks to us depends on its size and also on how far away it is. The Moon and the Sun look just about the same size, but the Sun is about 200 times bigger in diameter. It looks the same size as the Moon only because the Sun is about 200 times farther away.

But the Moon's size in the sky is not the only thing that would change if the Moon were much closer. Have you ever been at the seashore and noticed how the ocean comes in and goes out twice each day?

What if the Moon and Earth were as close as New York and California?

New York is about 3,000 miles (4,800 km) from California. If the Moon were this close to the Earth, it would be completely torn apart by the Earth's gravity. How come?

Just as the Moon's gravity pulls on the Earth, the Earth's gravity pulls back on the Moon—that's what keeps the Moon in its orbit. The Earth's gravity also has another effect. It makes tides on the Moon, even though the Moon doesn't have any water at all! It seems strange to talk about tides when there's no water. But the pull of the Earth's gravity is different on different parts of the Moon. So the different forces tend to pull the Moon apart. These are called **tidal forces.** The closer the Moon came to the Earth, the stronger the tidal forces would be, until the Moon was completely torn apart.

All that would be left of the Moon would be a lot of rocks and dust. These rocks and dust would spread out and orbit the Earth in a ring, like the rings around the planet **Saturn.** Scientists believe that could be how Saturn got its rings—from a moon that got too close and was torn apart.

We call these ocean changes the **tides.** The pull of the Moon's gravity on the Earth is the main cause of the tides. The Moon pulls a little harder on the parts of the Earth closer to it and a little less on the parts of the Earth farther away. All these different pulls make water pile up (change its level) at certain places, and the tides are the result.

If the Moon were much closer, its gravity would be much stronger, and the tides on Earth would be much more extreme. At high tide, the water could come many miles onto the land and flood areas near the oceans. At low tide, the water could go miles out to sea. All this would happen twice each day. You wouldn't want to be anywhere near the water's edge when the tide started coming in!

What if there were no Moon?

Of all the nine planets, only two—**Mercury** and **Venus**—don't have moons. If Earth had no Moon, we wouldn't have any tides. The ocean waters wouldn't come in and out each day. There would only be a very narrow strip of wet land right next to the edge of the oceans. With only this thin dividing line between ocean and land, life on Earth might have developed very differently.

Life on Earth originally developed in the oceans. Only later did ocean creatures come onto the land and develop into animals. These early land animals probably lived like crabs, spending much of their lives in the area at the edge of the oceans, where the tides came in and went out each day. In this way early land animals slowly got used to the land, but still spent most of their time in the water.

With much smaller tides, creatures living in the oceans would have had a much tougher

LUNAR LUNATICS. Ever heard of people or animals howling at the full Moon? A full Moon happens when the side of the Moon that faces the Earth is completely lit by the Sun. The full Moon was always believed to bring out people's emotions. People were once called "lunatics" (the Spanish word for Moon is "luna") because it was believed that they were driven mad by the moon. Each month check the calendar to find out when there is a full Moon. Do you feel different on those days? ◄

time adapting to the land. So without the Moon, all life might still be living in its original home—the sea.

THE MOON AND THE APPLE.

THE MOON AND THE APPLE.   There is a story that the scientist Isaac Newton discovered his **law of gravity** by seeing an apple fall from a tree while he was watching the Moon. Newton asked why the Moon doesn't fall from the sky, the way an apple falls from a tree. Newton thought it was because the pull of the Earth's gravity got weaker the farther away something was from Earth. He asked himself how *much* weaker the Earth's gravity would have to be for the Moon to stay in its orbit. And then he assumed that the same law of gravity applied to any two objects in the universe. What Newton discovered was that if the distance between two objects becomes twice as great, the force of gravity between them is $2 \times 2 = 4$ times weaker. And if the distance becomes 3 times as great, gravity is $3 \times 3 = 9$ times weaker. Newton's law of gravity might never have been discovered if the Earth didn't have a Moon.

If the law of gravity had not been discovered, people might never have thought about launching satellites. Who would think to launch a satellite if the Earth didn't have a natural one, the Moon? We also might never have traveled into space had we not had the Moon. The Moon is our nearest neighbor. For example, the Moon is 200 times closer to Earth than the planet Mars is, so a trip to the Moon is a much easier journey. Without a Moon, we would probably never have dreamed of going into space. ◀

What if you moved to the Moon?

Y ou'd step off the spaceship and leap for joy. It would be easy to leap high on the moon because its gravity is only one-sixth as much as Earth's. That means you'd weigh only one-sixth as much on the Moon. Because of the Moon's smaller size, it has less gravity than Earth.

You would look around and everything would be completely still. You could clap your hands and your friend wouldn't hear you, because there is no air to carry sound waves. The sky would be black all around you, all day and all night, because the Moon has no atmosphere. (See page 52, What if the sky were black during the daytime?) Why doesn't the Moon have an atmosphere? In an atmosphere, gas molecules move around every which way. Some of these molecules can leave the atmosphere and go into space if the

gravity pulling them down is too weak. The Earth's gravity is strong enough to keep oxygen and other molecules from drifting off. If the Moon ever had an atmosphere, all of it escaped into space because of the low gravity. With no atmosphere on the Moon, there is no oxygen to breathe. You could breathe in a space suit or in a completely sealed apartment, but you'd have to rely on oxygen supplies brought from Earth.

The space suit would also

WHAT GOES UP MUST GO . . . HUH?

If you release a helium balloon on Earth, it goes up because it is lighter than the air. What do you think would happen if you released a helium balloon on the Moon? *Hint:* There is no air on the Moon.

Answer: The helium balloon would fall. But if the helium gas were let out of the balloon, the individual helium molecules would escape into space because of their rapid motion. ◀

have to protect you against big temperature changes between day and night. During the day it can get to 264°F (130°C), like the warm setting in your oven. After the lunar surface is warmed by sunlight during the day, the nighttime temperature drops to –270°F (–170°C). The big temperature drop happens because there's no atmosphere to keep the heat from escaping into space.

You can see that life on the big ball of rock we call the Moon would be pretty rough. So why should humans ever live there when it's so comfy and cozy and normal here? Well, they might live on the Moon in order to use it as a jumping-off point into space.

Because of the Moon's low gravity, it would be a lot easier to launch a rocket from the Moon than it would be to launch one from the Earth. When we start making a lot of trips to the other planets, it might make sense to use the Moon, rather than the Earth, as a launching point. Maybe you'll be a space explorer of the future. Before you visit the planets, you could look up at the sky from the Moon and see the Earth. You might want to wave to your friends back home.

Nonstop Flight to Neptune
Visiting Other Planets

The solar system includes the Sun and everything in orbit about it. The nine planets and their moons all orbit the Sun, but moons orbit the planets as they go around the Sun. The solar system also includes many comets and asteroids (bodies smaller than planets that orbit the Sun). The nine planets, in order of increasing distance from the Sun, are Mercury, Venus, Earth, Mars, Jupiter, Saturn, Uranus, Neptune/Pluto. (The slash means that sometimes Neptune and sometimes Pluto is farther away from the Sun.) Some of the planets can be easily seen in the night sky, but they look just like stars. We can tell that they're

planets and not stars, because as planets orbit the Sun, they appear to move from night to night relative to the stars.

Humans have sent space probes to the planets, and the probes have sent back pictures. But no human has yet visited any of the planets. Some of the planets have too harsh an environment for life to exist on them. The planets nearest to the Sun (Mercury and Venus) are very hot, and the planets farthest from the Sun (Uranus, Neptune, and Pluto) are very cold. It is unlikely that people will ever try to land on any of these planets. But we might visit them, and either orbit them or visit one of their moons if they have any. Our neighboring planet Mars will probably be the first one that people will visit. No one knows when we will send people to other planets. It depends on whether we can develop the technology and also on how badly we want to go.

What if you visited Mercury?

Imagine waking up on Mercury, the closest planet to the Sun. It's sunrise. Mercury's sky is filled with a giant Sun—three times as big as the Sun looks from Earth. The sky stays black after the Sun rises (Mercury has no atmosphere—see page 52, What if the sky were black during the daytime?). But Mercury itself becomes too bright to look at without dark sunglasses. The temperature starts to climb quickly. From 50 to 100 to 500° F. It's soon hot enough on Mercury to bake a pizza. The temperature during Mercury's six-month-long day will eventually climb to 662° F (350° C), hot enough to melt lead. But of course, you yourself probably would have been baked by then.

It makes sense that Mercury is extremely hot, since it's the closest planet to the Sun (it's three times closer to the Sun than Earth). But, because Mercury lacks an atmosphere to keep the heat in, the nighttime temperature drops to –274° F (–170° C). This is far colder than anyplace on Earth. You'd be frozen in a matter of seconds without a space suit.

There's a bigger day-night temperature difference on Mercury than on any other place in the solar system. This is because Mercury rotates very slowly on its axis. It takes 176 days from one sunrise to the next, compared to one day for Earth. That means that each day and each night on Mercury lasts half that time—88 days. So temperatures would drop tremendously during each night, when there's no Sun in the sky. Then they'd rise tremendously for another 88 days, the planet baking under the Sun.

The only place that you'd be able to visit without freezing or frying would be in that narrow strip between the blazing hot days and the freezing nights. But because the planet rotates, the strip moves around the planet. You'd have to keep chasing after it—traveling as much as 54 miles (86 km) each day!

Let's suppose you were wearing a heat- and cold-proof space suit so you could go to Mercury. Look around. You wouldn't see a moon because Mercury doesn't have one. You'd see a black starry sky day or night, just like on our Moon. In fact, in many ways it would seem like living on the Moon. Mercury's gravity isn't strong enough to keep an atmosphere from escaping into space. On Mercury, you'd weigh less than half what you do on Earth.

The surface of the planet would look very similar to the surface of the Moon, with lots of craters. (**Craters** are circular depressions, or holes, caused by the impact of a meteoroid.) No water exists on Mercury. If it ever had any, the water would have long ago boiled away to space.

Hmm, maybe it would be best just to watch Mercury from our own planet, rather than visiting there. But even watching is not easy to do. Mercury is so close to the Sun that much of the time it's either directly in front of the Sun or behind it, as seen from Earth. Then the only time it can be seen is shortly before sunrise or shortly after sunset, when the Sun is below the horizon but Mercury is slightly above it.

What if you visited Venus

You'd be fried, squashed, poisoned, and **corroded** (eaten away by acid). Venus, the second planet from the Sun, is probably the worst place in the solar system to set foot on.

It's probably not worth trying to visit Venus, because condi-tions there are so hostile to life. But if people wanted to take the trouble, they might actually be able to build a spaceship that could land them on the planet. When you entered Venus's atmosphere, it might look like you're going through a giant ball of yellow cotton candy. The planet is completely covered by a thick layer of deadly yellow clouds made of carbon dioxide and sulfuric acid.

The clouds trap heat, acting like a blanket over the planet. This keeps the surface tempera-ture on Venus at a toasty 900° F (around 480° C)—even hotter than Mercury. The atmosphere on Venus is much thicker than Earth's. If you stood on the sur-face of Venus, the atmosphere would squash you instantly—pressing down on you with 90 times the pressure of Earth's atmosphere. **Atmospheric pressure** is the weight of all the gas in the atmosphere that lies directly above a square that is 1 inch (or 1 meter) on a side.

If you dared to breathe on Venus, you'd start to choke on the carbon dioxide atmosphere.

A TOPSY-TURVY PLANET. If you could stay on Venus, you'd soon notice something pretty weird. A day on Venus is longer than a year on Earth. Venus rotates *very* slowly on its axis, taking 243 days to complete one rotation. So a day on Venus is 243 Earth-days long. But Venus takes only 225 days to orbit or to go around the Sun, so its year is shorter than its day. When the next day finally comes, you wouldn't look for the Sun to rise in the east. Venus actually rotates backward (opposite to its revolution around the Sun), so the Sun would rise in the west and set in the east. ◀

"Kremlak, look! It's a UFO!"

Then you'd be pelted by drops of sulfuric acid. This acid is what makes the clouds yellowish. It is so destructive that it can dissolve most rocks and many metals—possibly including your spaceship.

Although Venus doesn't have any water or any life, it does have an amazing landscape—that is, if you could see it through the thick clouds. There are mountains steeper than any on Earth, and there are deep valleys. The tallest mountain is named Maxwell, and it's 7 miles (11 km) high.

Well, visiting Venus would be the pits. Still, it does look pretty and bright when seen from Earth. Look for it low in the sky shortly after sunset or before sunrise. For this reason, Venus is sometimes called the morning star or the evening star. Of course, it's a planet, not a star. Planets shine only because they reflect light from the Sun. Venus sometimes looks so bright that a lot of people have thought it was an unidentified flying object, or UFO.

What if you visited Mars ?

You'd want to bring your pick and shovel. You just might dig up some fossils of ancient Martian life. Unmanned spacecraft have gone to Mars and found no evidence of life. But recently, by studying a meteorite that reached Earth from Mars, scientists have found signs that life may have once existed on Mars. No, they didn't find skeletons of little green men. They found what looks like fossils of microscopic creatures. But some scientists aren't sure that the meteorite contains real fossils, and not something else that just looks like fossils.

We would be least surprised to find that, of all the planets, Mars once supported life forms. The fourth planet from the Sun, Mars is the most Earth-like of all the planets. It rotates on its axis once every 24 hours, just like Earth. Mars also has seasons, because its axis is tilted by the

same amount as Earth's. The planet even has ice caps at its north and south poles, just like Earth.

Before you pack your bags for a visit, you should know that Mars is also different from Earth in many ways. Some ways aren't so pleasant. Mars takes longer to orbit the Sun, so a year on Mars is two Earth-years long. Okay, so you could live with a two-year year. You can even live with its summer temperatures. A Martian summer is a comfortable 70° F (20° C).

But you wouldn't want to visit Mars during its winter. In winter, Mars is a gut-freezing –220° F (–140° C). The reason for the big summer-winter difference is the atmosphere on Mars. It is very thin and dry, and it doesn't hold much heat. The atmosphere is mainly carbon dioxide, and it also has a lot of fine red dust. The red dust makes the Martian sky look a pretty pinkish orange color. Sometimes fierce winds blow the fine red dust all across the planet.

It never rains on Mars, because there is no water on the surface. But there are plenty of signs, such as places that look like dried-up riverbeds, that the planet may once have had water. Mars might still have water frozen under its surface. And if there is trapped water, could there be some kind of frozen life? We just don't know yet.

ONE POTATO, TWO POTATO: "SYNCHRONOUS" ORBITS.   Mars has two tiny moons that orbit it. If you looked at the moons through a telescope on the Martian surface, you'd think you were seeing two giant potatoes. Gravity makes any big planet or moon round. But because of the moons' small size, they're not round. They're kind of lumpy and, well, potato-shaped.

One of Mars' two moons orbits the planet in almost the same time as the planet rotates once on its axis. This means that if you were standing on the surface of Mars, one moon would appear to always be in the same place in the sky. This is called having a **synchronous orbit.** On Earth, we launch some man-made satellites into synchronous orbits so that they can stay above a fixed point on the Earth. How long should it take a synchronous satellite to orbit the Earth? (*Hint:* How long does it take the Earth to rotate once?)

Answer: 24 hours. ◀

What if you visited Jupiter

Jupiter is the biggest of all the planets—more than twice the mass of all the others combined. If Jupiter were hollow, it could contain over a thousand Earth-size planets. Its huge size means that gravity is much stronger than that on earth. On Jupiter you'd weigh over two and a half times what you weigh on Earth. Of course, that's if you could actually stand on the planet—which you couldn't, because it doesn't have a solid surface.

Jupiter is the fifth planet from the Sun. It's five times farther from the Sun than Earth is, so it receives very little sunlight. In fact, the temperature on the surface is a very cold $-166°$ F ($-110°$ C). Jupiter is made mainly of hydrogen and helium, gases that are also found on Earth. But Jupiter is so cold that these gases are liquids inside the planet.

Only the innermost part of Jupiter (its core) is a solid, probably made of rock and ice.

Strangely, the planet really doesn't have a surface. There's no sharp boundary between the liquid inside the planet and the gases of its atmosphere above it. A spaceship trying to "land" on Jupiter would just sink into an atmosphere that got thicker and thicker until it gradually became a liquid—the thick atmosphere that is mainly made of hydrogen and helium. These very light gases would drift off into space on a planet with less gravity that was closer to the Sun, but not on chilly Jupiter.

On your way to Jupiter, you'd also be able to study Jupiter's giant red spot, which can easily be seen through a telescope on Earth. This spot is a giant storm that has lasted at least 300 years. This storm—a furiously swirling mass of gas—is bigger than the whole Earth. You'd also notice that the planet seems not quite round. The planet rotates on its axis so fast—once every 10 hours—that it flattens out a

Good Morning, Solar System!!!

little, sort of like when a pizza-maker spins a pizza and throws it up into the air.

As your spaceship nears Jupiter, you might instead be

Greetings from Jupiter.

Approaching Jupiter, you'd receive a welcome—radio signals being transmitted from the planet. These aren't broadcasts from any radio station there. Nope, it's not from station WIJU (What If Jupiter . . .). They're created by Jupiter's magnetism, which is 19,000 times stronger than Earth's. The radio signals are produced when tiny particles from space called **cosmic rays** pass nearby Jupiter. When this happens, the planet's magnetism makes these particles travel in spirals, and they give off radio signals. ◀

tempted to drop by one of the planet's 16 moons. Four of these are giants (for moons) and can easily be seen through a tele-scope on Earth. One is almost half as big as the planet Mercury. The moon called Io might make a particularly interesting choice for a visit. It's a little bigger than the Earth's Moon, and it's one of only two places besides Earth to have active volcanoes. Io looks very much like a pizza with all its colors. Don't take a bite!

What if you visited Saturn?

If you stopped by Jupiter before you visited Saturn, you'd have a good idea what was in store for you on Saturn. Saturn's almost as big as Jupiter, with a diameter 9 times the Earth's. It's also made of the same material as Jupiter—mainly hydrogen and helium. It's another really cold place because it's 10 times farther from the Sun than we are. And it spins extremely quickly on its axis—about once every 10 hours.

But Saturn has something that none of the other planets have—beautiful rings around it. They make people oooh and aah when they see Saturn through a telescope. Uranus and Neptune also have rings, but none that are as big and spectacular as those around Saturn. Saturn's rings are in a wide band extending out for 46,000 miles (74,000 km).

If you view Saturn through a telescope, you might not always notice its rings some of the time. That's because the rings are tilted. It's like looking at somebody wearing a hat with a big wide brim. If you look at the brim straight on, you might not see it.

Up close, Saturn's rings look completely different. They are not solid, but instead are made of lots of separate chunks of ice and rocks, all in orbit about the planet. Saturn's rings are probably the material left over from a

▼

PLANET OR MOON? Even though Titan is bigger than the planet Mercury, it is a moon, not a planet. Planets orbit the Sun, and moons orbit planets. What would you call something that orbited a moon? Maybe a moonlet? ◀

moon that got too close to the planet and was torn apart by its gravity.

Maybe the rings are the remains of Saturn's 21st moon. Saturn has 20 moons, and one of them, Titan, is huge. It's bigger than the planet Mercury. In fact, Titan's gravity is able to hold down an atmosphere. It's the only moon in the solar system that has an atmosphere.

Saturn's rapid spin creates enormous winds of over 1,000 miles per hour (about 1,600 kph) on its surface. But Saturn, like Jupiter, really doesn't have a surface. There's no sharp boundary between the liquid making up the planet and the atmosphere above it. Saturn's atmosphere of hydrogen and helium is 30 times thicker than the one on Earth. About the only thing Earth-like about Saturn is its gravity. Bigger planets usually have stronger gravity, but the material a planet is made of also affects the strength of gravity. Even though Saturn is much bigger than Earth, it's made of such light material that its gravity is the same as Earth's. In fact, Saturn is so light that if you had a big enough pool of water, Saturn would float in it!

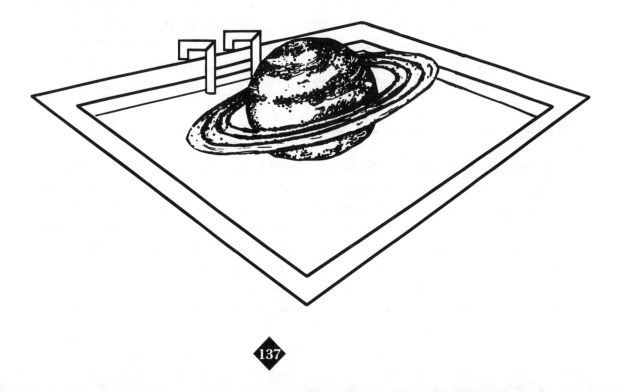

What if you visited Uranus or Neptune

The farther out you go in the solar system, the colder it gets. You probably wouldn't want to visit chilly Uranus or Neptune, where the temperature averages about 357 degrees F below zero (–216° C).

These two big blue planets are both so far from the Sun that they look like faint stars from Earth. In fact, until around 200 years ago, we didn't even know they were planets. Uranus was discovered when someone said "Hey, this star is moving." Neptune was discovered when someone else said "There's something really strange about the way Uranus is moving." It seemed like something was tugging on Uranus—maybe the gravity of another planet. Scientists soon located Neptune in the direction of the pull.

At first glance, Neptune and Uranus could be twins. Both are about four times bigger than the Earth. Both are made of the same material—mainly liquid and gaseous hydrogen. They each rotate on their axis once every 17 hours. Both planets also have rings around them. But unlike Saturn's rings, their rings are very thin and dark and can't be seen from Earth.

THE MOON LOOKS JUST SMASHING TONIGHT! Both planets have lots of moons—Uranus has at least 15, and Neptune has at least 8. Triton, one of Neptune's moons, orbits the planet backward—in the opposite direction that the planet spins. Due to its backward orbit, Triton is pulled by Neptune's gravity a tiny distance closer to the planet with each orbit. Eventually, Triton will come too close to Neptune and will be torn apart by its gravity. The pieces of the moon will then spread out and form a big ring around Neptune. Then Neptune will look much like Saturn looks now. ◄

OUCH! I could use an aspirin the size of Pluto!

Yet the two planets have some wacky differences. Neptune has a great dark spot on it just like Jupiter does. The dark spot, as big as the whole Earth, is a giant swirling storm with winds of around 1,500 miles per hour (2,400 kph).

Uranus is the only planet in the solar system with an axis tipped all the way over on its side. At one time of its year, its north pole faces the Sun, and half an orbit later (that's 42 Earth-years later!), it points away from the Sun. How did Uranus get tipped over? Nobody knows for sure, but it could have been the result of a big collision. When the solar system was very young, chunks of planets smashed into one another sort of like bumper cars at a carnival.

What if you visited Pluto

If you ever got to Pluto, you'd have to say good-bye to the Sun. Seen from this rocky, icy planet, the Sun would look like any other bright star. Not surprisingly, it's c-c-c-cold on Pluto (−369° F, or 223° C).

You probably think of Pluto as being the farthest planet from the Sun, but it's not always so. Depending on when you're reading this, Neptune might be the farthest from the Sun of all the planets. (Neptune is farther from the Sun than Pluto between the years 1979 and 1999.) Here's why Pluto is sometimes closer than Neptune to the Sun. Pluto's orbit around the Sun is more elliptical (oval shaped) than any other planet. Because of the shape of its orbit, the distance from Pluto to the Sun changes greatly. It takes Pluto 250 years to orbit the Sun. For part of that time it is actually closer than Neptune to the Sun.

Pluto is a small planet made up of rocks and ice. Its gravity is the weakest of all the planets. You'd weigh only about one-sixteenth as much on Pluto as you weigh on Earth. So you'd be able to jump sixteen times higher. Pluto very closely resembles one of the moons of Uranus or Neptune. In fact, Pluto may be a moon of Neptune that got away.

Pluto does have one moon of its own, Charon—and what a moon it is! Charon is bigger than half the size of Pluto itself. With a moon this big, Pluto is more like a double planet, because the moon and planet orbit each other. As the moon and planet move around each other, each keeps the same face toward the other. If you were standing on Pluto and looking up at its moon, you would see that its moon never rises or sets, but always stays at the same point in the sky.

Pluto is the newest planet to be discovered. As seen from Earth, it looks like a very faint star. For that reason, it was not discovered until around 70 years ago. Could there be planets beyond Pluto that we still don't know about? Sure. But they'd have to be either very small or very far away. Otherwise we'd have seen them by now.

The Mystery of the Double Orbits *How to see double planets.* Make two equal-size clay balls. Stick one ball on each end of a pencil. Balance the pencil on your finger, and mark the point on the pencil directly above your finger (the balance point). The balance point should be exactly halfway between the balls, at the middle of the pencil. Tie a string on the pencil at the balance point, and let the pencil and clay balls hang from the string. Give the pencil a spin around the string so that the two clay balls go in a circle. If the clay balls are the same size, each one goes in the same size circle as the pencil spins—the balls orbit a point halfway between them. Break a big piece of clay off one of the balls, so that they are different sizes, and repeat the whole experiment. Now, the smaller ball goes in a big circle, and the bigger ball goes in a small circle. If one ball were much bigger than the other, it would hardly move at all during the orbit of the smaller one.

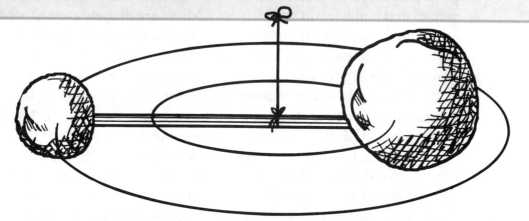

The large ball, which is closer to balance point, follows a smaller circle than the small ball, which is farther away from balance point.

The small ball travels a longer distance than the large ball. Because it travels farther in the same time, the small ball is moving faster.

What if you visited planets in another solar system

Our Sun is probably not the only star with planets in orbit around it. Scientists think that many stars have solar systems like our Sun's. If you wanted to be comfortable, you'd probably look for a solar system that had a world as much like Earth as possible. That might not be so easy, even if it were possible to get there.

You might find a solar system that had only two planets—a giant such as Jupiter, and a tiny one such as Pluto—but no medium-size world just like Earth. Or you might find a solar system that had an Earth-size world, but it was too close to the Sun and much too hot to live on. Or maybe the world was the right size and distance to the Sun but had an atmosphere you couldn't breathe.

How do scientists know that other stars have solar systems? No one has actually seen planets around other stars, even with the most powerful telescopes. Planets don't give off light like stars do. They only shine by the light they reflect. Planets orbiting our Sun are easy to see because they are a lot closer than other stars. But planets orbiting other stars are too faint to see in the glare of the light from their star.

Scientists use something called the **Doppler Effect** (an increase or decrease in wavelength due to motion) to find out if there are planets around other stars. Have you ever heard the change in pitch of a police or ambulance siren as it passes you? It sounds something like: e-e-e-e-e-o-o-o-o-o-o. Before the police car passes, the sound waves from its siren are all bunched together, and its wavelength (distance between wave crests) is shortened. After it passes, the waves are stretched out, and the wavelength is increased. These changes in wavelength of sound are heard as a changing pitch as the siren passes.

142

The Doppler Effect also happens with light waves. When a light source approaches, the light waves are bunched up. The decreased wavelength is seen as a change in color—the light is slightly bluer. If the light source moves away, the wavelength is slightly longer, and the light is slightly redder. These changes are much too small to be seen by eye. They can only be seen by using sensitive instruments called **spectrometers.**

E – E – E – O – O – O

THE DOPPLER EFFECT

Shorter wavelength equals higher pitch.

Longer wavelength equals lower pitch.

So how do scientists use the Doppler Effect to find planets around other stars? When a planet orbits its star, the star also goes in a small orbit. (See the Mystery of the Double Orbits experiment on page 141.) Suppose the star's orbit is seen on edge from Earth. For part of its orbit, the star moves toward Earth, and for part of its orbit, it moves away. This motion makes

the star appear slightly bluer and then slightly redder during each orbit. Scientists measure these small changes in the wavelength of starlight. Based on these measurements, scientists think that they may have found more than 12 stars that have planets in orbit about them. From these measurements scientists can figure out how big the planets are and how long they take to orbit their star, but not much else.

Beam Me Up

The Stars and Beyond

Our Sun, along with about 100 billion other stars, belongs to a group that we call the **Milky Way.** If we could see the Milky Way from the outside, it would look like an enormous disk with a slight bulge in the middle—sort of like an egg fried "over easy." Stars are only found in groups called **galaxies.** Some galaxies have a shape like the Milky Way (a **spiral galaxy**), whereas others, called **elliptical galaxies,** are shaped like an egg. Scientists are not sure how the galaxies were originally formed. All the stars you can see without a telescope (plus billions more you can't see) are in "our" Milky Way galaxy. Our galaxy is just one of billions of such giant collections of stars that fill the universe. The

universe is what we call everything that exists. It has no edge to it, because then it would have an inside and an outside. Galaxies other than the Milky Way are too far away from Earth for us to see individual stars in them without a powerful telescope. They look just like fuzzy patches of light in the sky. But each little fuzzy patch contains billions of stars—some of which may have solar systems and maybe intelligent life.

What if you traveled to the nearest star

There are a lot of problems with traveling to the stars. First, you can't land on them. It would be like trying to land on the Sun. They're too hot, and they're not solid either. You'd have to hope that the star had a planet that you could use as home base. Otherwise, you'd just spend your time orbiting the star once you got there.

But you can dream about it. The nearest star to Earth is called **Alpha Centauri.** Even this is so far away that scientists use a special unit to measure its distance: the light-year. A **light-year** is the distance that light travels in a year, which is about 6 trillion miles (10 trillion km). A light-year is so far that you'd have to travel around the world 236 million times to go 1 light-year. Alpha Centauri is over 4 light-years away from us.

"Are we there yet?"

FAR-**O**UT **S**TARS!   Here are two ways to get an idea just how far away the nearest star is. First, suppose that the Sun and the nearest star were separated by the length of a football field. On that scale, the Earth would be only 1/16 inch (1.6 mm) away from the Sun—about the width of the lead in your pencil. No wonder our Sun looks so much bigger and brighter to us than any other star in the sky.

Now, here's how former astronaut Michael Collins thinks about how far away the stars are. He says to go into the backyard and point a flashlight beam at Betelgeuse (pronounced "beetle juice"!). That's a bright star in the constellation of Orion. (**Constellations** are pictures or patterns made by groups of stars when seen from Earth. Some people think Orion is a picture of a hunter.) Your light has to travel hundreds of millions of miles through space to reach the star. In the couple of hundred years it takes the light to get there, you won't be around. Perhaps your great-great-great-great-great-great grandchild will be alive when your light beam reaches Betelgeuse. ◀

With today's spaceships, even a trip to the nearest star would just take too long. For example, astronauts were able to travel to the Moon in a couple of days. But the trip to the nearest star is a 100 million times farther than the Moon. It might take astronauts a million years to get there using today's spaceships. There wouldn't be too many volunteers for that trip!

Even if spaceships could travel almost at the speed of light, a trip to the nearest star and back would take over eight years. Would you want to spend eight years of your life cooped up in a spaceship? (Actually you wouldn't have to. The trip might last eight years according to people on Earth, but it could be a lot less for people in the ship. That's one of the ideas of Einstein's theory of relativity. See page 56, What if light traveled very slowly?)

What if you traveled to the center of our galaxy?

If thinking about a trip to the nearest star wore you out, think about a trip to the center of our galaxy—that's around 25 million times farther away!

The Milky Way galaxy is a collection of around 100 billion stars that includes our solar system. Without a telescope, we can see only a tiny fraction of these 100 billion stars. The rest are just too faint and too far away to see.

In fact, every one of the 2,000 or so stars you can see on a clear night is in the Milky Way galaxy. We can't actually see the shape of our galaxy (sort of like a fried egg, with a bulge in the middle), because we're inside it. Our Sun is located about a third of the

HOW DO WE KNOW THAT WE'RE INSIDE A GALAXY? Because the galaxy has a flattened shape, and because we're inside, it looks different to us when we look in different directions. For example, look at the sky on a very dark night out in the country. In certain directions you may be able to see a very faint milky band across the sky. That band is why we call our galaxy the Milky Way. It's actually an edge-on view of one part of the galaxy. The milky appearance is from the many stars that you are looking through.

If you could travel out to space where the milky band is, you'd be surprised how empty the sky would look. Even though the stars look close together from here, most of the galaxy is empty space. If you took 12 tennis balls and spread them out across the United States, they would be more crowded than the stars in our galaxy! ◄

way in from the edge. The Sun and all the other stars in the galaxy very slowly rotate about the center. That rotation is what makes the galaxy flattened. Just think how a pizza gets flattened when a pizza-maker gives it a spin and tosses it up into the air.

You might be wondering how come a galaxy can get flattened when it rotates very slowly, but a pizza needs to rotate fast before it gets flattened. That's because galaxies are much bigger than pizzas. Most of a galaxy is very far from the axis as it rotates. The parts of a rotating object that are farther from the axis fly outward more easily. So a big object such as a galaxy doesn't have to rotate so fast before parts of it start flying

outward. (See the Round and Round She Goes experiment on page 14.)

If we ever got to the center of the Milky Way galaxy, things would look very different. There are many more stars in the center of the galaxy than out where we are. And the stars in the center are much closer together. We also might find a giant black hole in the center of the galaxy. Stay away from that!

♥ HOME SWEET HOME ♥

149

What if you traveled to another galaxy

And you thought the center of our galaxy would be a long trip! Other galaxies are even farther away from us. Yet, amazingly, they're in our same neighborhood. Our neighborhood of galaxies is called the **local group.** Our nearest neighbor galaxy is called **Andromeda.**

On a clear night Andromeda will appear as a fuzzy patch in the sky. Yet Andromeda is so far away that the light you see from it started on its journey over 2 million years ago. That makes the distance to Andromeda about 60 times the distance to the center of our own Milky Way galaxy.

Andromeda is a galaxy much like our own. It contains around 300 billion stars. If you were to count one star per second nonstop, it would take you more than 9,000 years to count all the stars in that galaxy. We probably will have sent spaceships to Andromeda by then!

HOW DO WE KNOW ABOUT DARK MATTER? Most objects in the universe are too far apart to bump into one another too often. But because of gravity, they pull on each other, no matter how far away they are. Even when you can't see something, you can tell it's there by the effect of its gravity on another object. That's how we can tell some other stars have planets: as the unseen planets orbit a star, their gravity pulls on it, making the star go in a small orbit. Galaxies move through space, and their gravity changes the motion of other galaxies. But there's not enough matter in the galaxies to make them move the way they do. Something else must be pulling on them. We call that something else dark matter. ◀

"I don't know what it is. You don't know what it is. Nobody knows what it is. But if we don't back up and change course, we'll bump into it again."

Let's imagine you could travel to another galaxy. You're in a spaceship traveling to Andromeda. Once you left our galaxy, the sky would be incredibly black. That's because the only place visible stars are found is in galaxies. None exist in the space between them.

Whoa, but what's this? Your imaginary spaceship just bumped into something huge. Beware: the space between galaxies is not really empty.

Scientists think there's a lot of "stuff" out there that they call **dark matter.** They call it dark matter because we can't see it, and nobody knows for sure what the stuff is. (See the box on dark matter to find out why they think it's there.) It could be stars that don't shine brightly enough for us to see them, or maybe stars that have already burnt out.

What if you went to the edge of the universe

Were you thinking it would take a really, really, really long time? Actually, it's a trick question. The universe includes everything we can possibly see in the sky—all the planets, stars, and galaxies as well as many things we can't see. There is no edge of the universe. Anywhere you went in the universe, things would look pretty much the same. You'd see galaxies—giant groups of stars— headed away from you in all directions.

Why would the galaxies all be moving away? The galaxies in the universe are flying apart from each other from a gigantic explosion that occurred about 15 billion years ago. That explosion, which is believed to have given birth to the universe, is called the big bang. One way to picture this is to think of the galaxies as dots on a giant balloon. The balloon is the universe, and it keeps getting bigger and bigger. Picture yourself sitting on one dot. As the balloon gets bigger, you'd see all the other dots moving farther away from you.

When scientists look farther and farther into space with telescopes, they see more and more galaxies. They can tell that most of these galaxies are moving away from us, even though they don't see any movement. How come? When galaxies move away from us, the light waves we see from them are stretched out, and this makes their color shift toward red. This so-called **red shift** is an example of the Doppler Effect we talked about earlier on page 142. The red shift tells us that galaxies are moving away from us. The farther away a galaxy is, the faster it is moving away—so the greater its red shift.

152

The time it takes light to reach Earth from distant stars and galaxies depends on how far away they are. Light from other galaxies takes millions or billions of years to reach us. The farther out we look in space, the earlier the light had to start out to reach us now. So by looking out in space, we're really looking back in time.

The most distant objects in space we could possibly see would be 15 billion light-years away. The light from such objects travels a distance of one light-year every year, and it would have started out around 15 billion years ago. If we could see that far out, we'd be looking back in time at the birth of our universe—the time of the big bang.

What if there are intelligent aliens

Guess what? There might be. Some astronomers believe that there are at least a trillion planets in the universe that could support life. (A **trillion** is a thousand billion or 1,000,000,000,000.) Is it reasonable to think that our cozy little Earth is the only one of these trillion planets with intelligent life? Probably not. Most scientists think that under the right conditions, life would develop by itself.

Because the possibility is so tempting, scientists have already tried to listen in on aliens. They use giant radio telescopes to see if there are any alien radio signals coming from distant stars or planets. Radio telescopes look nothing like ordinary telescopes. They look very similar to satellite TV dishes, and they work in exactly the same way. Radio and TV signals can go through space. And even though radio signals get weaker after traveling a long distance, they can be picked up by powerful radio telescopes.

But that's not such an easy thing to do. One problem is that we now get natural radio signals from space. In fact, several of the planets, including Jupiter, give off radio signals, and nobody thinks that's a sign of life. How can they tell when a radio signal is natural, and when it's broadcast by aliens?

A scientist named Frank Drake has spent years waiting for aliens to signal him. During his Project Ozma, he tuned in to over 600 stars. If aliens are trying to contact us, he figured, they wouldn't just say "Howdy" and leave it at that. He listened for a pattern of signals, repeated over and over again. If dashes are long signals and dots are short ones, an alien message might go something like "dot-dash-dot-pause-dot-dash-dot," which is kind of like a ship's cry for help, S.O.S—Save Our Ship (or Save Our Spaceship!). But you'd probably want to hear a more complicated message than that before you really thought it was an alien signal.

So far, no scientist, including Drake, has picked up an alien message. But Drake decided, hey, why sit around waiting for an alien to phone us. Why not send our own message? Drake's radio message consisted of 1678 dots and dashes. If the aliens are really smart, they'll put the signals together to make a picture showing all the important things about Earth life—from the structure of atoms to the human figure. It will arrive in Messier (a cluster of stars in our galaxy) by the year 25,000. We'll probably have shaken hands with aliens by then!

"Hey! *Hey,* Buddy!!"

155

What if tiny aliens landed on Earth

We might not even notice them. The next time you see a strange-looking insect, you might want to think about where it might have come from. (Just kidding!)

Usually, when we think of aliens landing on Earth, we imagine them as being about the same size as we are. But what if aliens the size of insects landed on Earth? Could aliens the size of insects find their way to Earth? The aliens would need brains to develop an advanced technology that includes spaceships. It's hard to imagine insects that lack brains developing a complex technology. Could it be that on another planet what we call a brain would be the size of a pinpoint? After all, scientists can make very tiny computer chips. But, of course, a computer *chip* is not the same as a computer. The complexity of a brain depends on the number of cells it contains, and cells must have some minimum size. So, a computer with the complexity of the human brain would not be so tiny. Insect-size brains, if they could exist, probably wouldn't be too smart!

What if aliens were microscopic?

That is, what if they were the size of germs—so tiny that they could only be seen under a microscope? These aliens couldn't be intelligent, and wouldn't come in spaceships, but maybe they would hitchhike a ride aboard a meteorite. Some scientists think that's how life on Earth actually began billions of years ago!

But suppose there were insect-size intelligent aliens. And what if we had an intergalactic war with them? We can't assume that we would win just because we're the big guys. Most likely, the tiny extraterrestrials would be way ahead of us in technology—otherwise they wouldn't have been able to get here. With their advanced technology the tiny aliens might be able to win a fight with us giants.

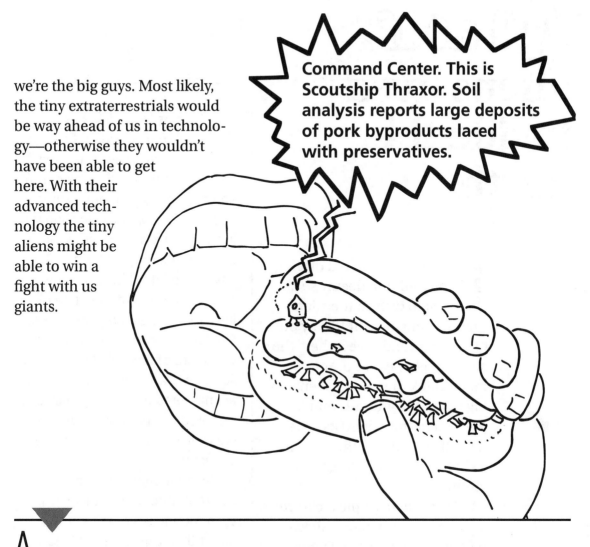

Command Center. This is Scoutship Thraxor. Soil analysis reports large deposits of pork byproducts laced with preservatives.

A TINY ALIEN TALL TALE. A science fiction story tells how aliens sent a message to Earth that they were going to be arriving at a certain place and time. The government arranged a big party to greet the aliens when they arrived. They even had marching bands and hot dog stands for the hungry extraterrestrials! As the aliens landed, they sent another radio message that they had just arrived. Yet, no Earth people at the landing site saw anything. They figured the aliens must have gotten lost. The aliens' final radio broadcast described what they saw last: being swallowed up in a huge cave, with tall grass all around. It turns out that the tiny alien ship had landed on somebody's hot dog (covered by sauerkraut), and was eaten. ◀

What if aliens lived inside the Earth?

People might avoid drilling oil wells too deeply so as not to disturb the aliens. And they might want to stay out of underground caves where the aliens might be found. You might think the idea of aliens living under the ground sounds pretty silly. But scientists have recently found that there really is a kind of alien life living underground!

These microscopic creatures are known as **extremophiles,** meaning that they like living under very extreme conditions, where no other kinds of life could exist—such as deep underground. No one believes that extremophiles are aliens from outer space. But extremophiles could live under the harsh conditions that can be found in space or on some of the planets.

THE TREE OF LIFE. Life has developed or evolved on Earth in much the same way that a tree grows—starting with a main trunk and then branching out as time goes on. The original life form was like the trunk of the tree. As life evolved, the original form of life developed into different forms. You might say that the "tree of life" split into separate branches. One branch includes all bacteria, which are single-celled organisms. Another branch includes all plants, animals, and fungi (fungi includes mushrooms and molds). And now we've found a third branch: the extremophiles.

Extremophiles are very common throughout the Earth, even though we didn't know about their existence until recently. In fact, scientists now think that the total weight of all these microscopic underground creatures is more than that of all the creatures living on the surface. ◄

Extremophiles have been found in hot water vents next to undersea volcanoes in water hotter than 600° F (316° C). That's way above the temperature at which water boils at sea level (212° F or 100° C). They have also been found 2 miles (3.2 km) under the ground in Virginia. Scientists once thought that unless conditions were just right, life could not develop on a planet. The presence of these creatures on Earth means that life might be found in a wider range of space environments than once thought. For example, Europa, one of Jupiter's moons, is believed to have an underground ocean. Maybe Europa's ocean is a very popular place for extremophiles to live.

Getting back to Earth, scientists think that extremophiles are so hardy that they can live even inside the molten lava inside undersea volcanoes. They can swim in acid and get their food and energy from the molten rocks. Extremophiles are so different from every other form of life on the planet that scientists now consider them to be a separate kind of life—as different from plants and animals as bacteria are.

What if aliens who looked just like us landed on Earth?

It's very unlikely that if a spaceship landed, creatures who looked, well, human would walk out. In the movies, aliens often look amazingly similar to people. Sometimes just small differences, like pointy ears or slimy skin, sets them apart. Yet if aliens actually do land some day, they would probably look very different from us.

Think about it. Creatures born on another planet would have adapted to life on that planet. For aliens to look just like humans, the aliens' home planet would have to be exactly like Earth. It's more likely that the aliens' home would be different in many ways and still support life. For example, it could be warmer or colder, have stronger or weaker gravity, or have a different kind of atmosphere. Each of these differences would have an effect on the kinds of creatures that could live on the alien planet.

For example, on a planet with strong gravity, aliens would probably be smaller than creatures on Earth. If they grew the same size as Earth creatures, they would need very thick legs to support their greater weight because of the strong gravity. They probably wouldn't be able

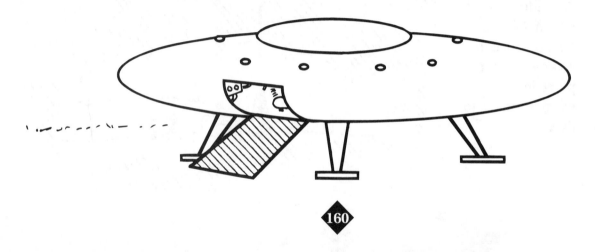

to get around too easily with such thick legs—unless they got around by rolling. On a very cold planet, most creatures might have very thick furry coats to keep them warm.

There's one way that aliens could look just like humans—it sounds like a *Star Trek* episode. Maybe all humans came from a group of aliens that left their home planet and landed on Earth a long time ago. Still, if that were too long ago, we probably would have developed differently from them. We wouldn't look the same anymore. Some scientists think that life actually began on Earth from germs or "seeds" that drifted from space. But even if all life on Earth did start from seeds from space, scientists believe that humans evolved from apes living on planet Earth—not from human-look-alike aliens!

What if the aliens' home planet was exactly like the Earth?

It's still unlikely that the aliens would look just like us. The appearance and abilities of creatures evolve, or develop, and change over millions of years. These changes help the creatures to survive and to better take advantage of their environment.

If you had two identical planets with the same creatures on them to start with, different things would happen to each planet. Maybe one would be hit by a giant meteorite. The planets would certainly not look the same millions of years later, and neither would the creatures on them.

What if aliens who wanted to talk only to dogs landed on Earth

In movies about *nice* aliens, the aliens always seem to know that people are the species on Earth worth talking to. Often the aliens even speak English! Maybe they received all the radio and TV programs we broadcast and learned the language on the way here. But what if aliens landed and wanted to talk only to dogs, not to humans? Would that mean that dogs, not humans, are the smartest species?

Not necessarily. It may just mean that aliens don't see our world the way we see it. When you see people walking dogs on the street, doesn't it sometimes seem as if the dogs are leading the people? After all, they run on ahead, dragging their masters behind. And their masters even clean up after them!

The point is, in order to even begin thinking about aliens, we have to stop thinking like humans. That's almost impossible because we can't be anything but human. But close your eyes, and then open them again. Look at the world as if you were from another planet and as if people were the aliens.

Think about what you'd see as you approached the Earth. You might see millions of cars streaming along the highways. Wouldn't it make sense to think of the cars as aliens? Maybe the people inside the cars would seem like the alien's brains. As your spaceship landed, all the cars would let out a strange beeping and honking noise. Maybe you'd think that the aliens were speaking to you!

Have aliens landed already?

Some people claim that aliens have already landed and that the government is covering it up. Other people claim that aliens have taken them aboard their spaceships and done all sorts of medical experiments on them. Most scientists do think that aliens really exist out in space. But they also think that people who tell such stories about aliens landing are just imagining things. The scientists say that they want more proof before they'll believe that aliens have really landed. What do you think?

Anyway, if aliens and dogs did start talking to each other, we humans would have quite a problem. We can't understand the language of dogs or aliens. Instead of dogs begging from us, we'd be begging the dogs to tell us what the aliens said. We'd probably start seriously studying dogs to learn their language. Woof!

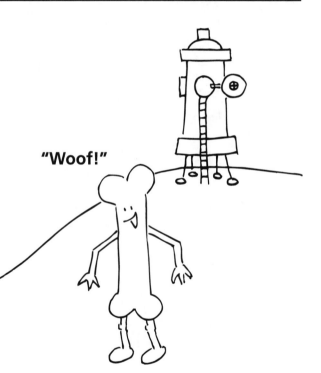

"Woof!"

Glossary

Alpha Centauri The nearest star to our Sun. It takes light 4.2 years to reach us from this star, so we say the star is 4.2 light-years away.

Andromeda A nearby galaxy much like our own Milky Way, only larger.

asteroid One of thousands of small, rocky, planetlike objects in orbit about the Sun. Most of the asteroids orbit the Sun between the orbits of Mars and Jupiter.

astronomy The science that studies everything in space beyond the Earth's atmosphere, including the planets and the stars.

atmosphere The layer of gas, such as air, surrounding a planet.

atmospheric pressure The weight of all the gas in the atmosphere that lies directly above a square 1 inch (or 1 meter) on a side.

atom The smallest particle of one of the basic substances (elements), which all other substances are made from.

atomic clock A clock that keeps track of time based on counting atomic vibrations.

attract Pull toward.

axis An imaginary line through a rotating object around which the object spins.

ball lightning Lightning in the form of a small glowing ball of light.

Bernoulli's Principle A principle of physics that connects reduced pressure to movement of air.

big bang A tremendous explosion that occurred around 15 billion years ago in which the universe was born.

billion A thousand million, written as a one with nine zeroes after it: 1,000,000,000.

biologist Scientist who studies living things.

biosphere The part of the Earth inhabited by living organisms.

Biosphere 2 A man-made enclosed environment that was intended to simulate the Earth's atmosphere and ecosystems.

black dwarf A burnt-out star at the end of its life.

black hole An object whose gravity is so strong that nothing, not even light, can escape from it; usually formed when a star has burnt out and collapsed from its own gravity.

boiling point The temperature at which a liquid boils.

bore To drill a hole.

calorie A unit of energy.

camouflaging Taking on colors or patterns to blend into the environment.

carbon dioxide A gas that is present in the atmosphere, and is absorbed by plants.

Celsius temperature Temperature measured on a scale where water freezes at 0 degrees and boils at 100 degrees; named for Anders Celsius (1701–1744), a Swedish astronomer.

cell The smallest unit of living matter.

chemist A scientist who studies the properties and reactions of matter.

climate The average weather over a period of time.

cloning Making identical copies of an organism from a single cell.

compass A device that uses a magnetic needle to determine direction.

constellations Pictures or patterns made from groups of stars as seen from Earth.

continuous Smooth, or all one piece.

copper A metal from which pennies are made.

core The central part of a sphere, such as a planet or star.

corroded Eaten away, by acid for example, or rusted.

cosmic rays Tiny particles bombarding the Earth from space.

Crab Nebula An expanding cloud of dust and gas created by a supernova in the year 1054.

crater A circular depression (hole) on a planet or moon caused by the impact of a meteoroid.

crest High point, on a wave, for example.

crust (of Earth) The outer layer of the Earth.

dark matter The invisible matter out in space that we know is there based on its gravitational pull.

diameter The length of a line from one side of a circle (or sphere) to the other, which passes through the center.

DNA The basic molecule of all living things that describes how they are made.

dominate Control.

Doppler Effect A change in wavelength due to motion.

double planet Two bodies in orbit about each other that orbit the Sun together.

double star A pair of stars in orbit about each other.

echo Sound reflection.

echolocation The process that bats and dolphins use to find the distance, size, and shape of objects, based on the echoes returned from a series of clicks or squeaks the creatures make.

eclipse The blocking of light from one heavenly body by another.

ecosystem A community of living organisms together with their environment.

electric current The flow (movement) of electrons.

electricity A form of energy produced by the movement of electrons.

electron A tiny particle usually found inside atoms.

ellipse An oval shape.

elliptical galaxy An egg-shaped galaxy

energy The ability to do work.

equator An imaginary circle going around the center of the Earth that is located an equal distance from the North Pole and the South Pole.

Europa One of the moons of Jupiter; believed to have an underground ocean.

evaporation The process whereby molecules leave a liquid surface and form vapor (gas).

evolve To develop and become better adapted to the environment.

extinct No longer existing.

extremophiles Recently discovered microscopic creatures that live under extreme conditions such as extremely high temperatures and pressures.

Fahrenheit temperature Temperature measured on a scale where water freezes at 32 degrees and boils at 212 degrees; named for Gabriel Daniel Fahrenheit (1686–1736), a Dutch scientist.

farsighted Unable to see close things clearly.

filament The wire that glows brightly inside a lightbulb.

fluid A form of matter that can flow. A gas or liquid.

food chain The sequence of animals that eat other animals.

force A push or a pull.

fossil The remains, an impression, or a trace of a plant or animal that lived a long time ago that has been preserved in the Earth's crust.

fossil fuel A fuel such as coal, oil, or gas formed from the remains of plants or animals that lived a long time ago.

friction The force created when objects rub against each other.

full moon When the moon's sunlit side faces the Earth.

fungi A form of life different from plants and animals that includes mushrooms, lichens, and molds. Fungi is plural of fungus.

galaxy An enormous grouping of stars.

gas A form of matter than can flow, but does not occupy a fixed volume.

genes The smallest parts of DNA that control any specific characteristic of the body.

genetic engineering Changing a gene on purpose, in order to change some characteristic of an organism.

geologists Scientists who study the structure and history of the Earth.

gravity A force of attraction that pulls matter to matter. The Earth's gravity pulls all things toward the center of the Earth.

hail Balls or lumps of ice that fall from the sky.

helium A gas that is lighter than air.

hemisphere Half of a sphere.

hibernate To spend the winter in an inactive state, like bears and some other animals do.

horizon Line where the sky meets the ground.

hydrogen The lightest element, which is the main substance found in the Sun.

infrared Waves like light, but having a somewhat longer wavelength, given off by objects or people that are warmer than their surroundings.

Io One of Jupiter's moons.

Jupiter Fifth planet from the Sun.

kinetic energy The energy of a moving object.

law of gravity Sir Isaac Newton's law that says that a force called gravity pulls all objects together. The strength of the pull is affected by the mass of each object and how far apart they are.

leap year A year with 366 days. February has 29 days rather than 28; occurs every four years.

lens Curved transparent object that creates an image by focusing light.

light-year The distance that light travels in one year.

liquid A form of matter that can flow and that takes the shape of its container.

local group Our galaxy and about 20 other neighboring galaxies.

lunar Referring to the Moon.

mammals All the species of animals that are warm-blooded, have hair, and breast-feed their young.

Mars Fourth planet from the Sun.

mass The amount of matter in an object.

mass extinction The extinction of a large number of species of plants and animals at around the same time.

mate Joining together of a male and female to produce offspring.

Mercury The closest planet to the Sun.

meteor A rock from outer space that burns up when it enters the atmosphere; also called a shooting star.

meteorite A rock that falls out of the sky and lands on Earth.

meteoroid A rock in outer space.

microwave Light waves with a wavelength too long to be seen.

Milky Way The galaxy that contains our Sun; also, a faint milky looking band across the sky consisting of many faint stars.

molecule The smallest piece of a substance that retains the properties of the substance; a combination of two or more atoms.

mutation A change made in copying the DNA of a plant or animal. Most mutations are harmful, but some are helpful.

navigate Find the way.

nearsighted Unable to clearly see things that are far away.

Nemesis A possible dim companion star in orbit about our Sun that has not yet been seen.

Neptune Eighth planet from the Sun—but sometimes it's the ninth, depending on where Pluto is in its orbit.

nitrogen A gas found in the Earth's atmosphere.

orbit The path of one body around another, such as the orbit of the planets around the Sun.

organisms Living things.

oxygen The molecule contained in air that animals need to breathe.

pendulum Something swinging on the end of a string or a stick. Since each swing of the pendulum takes the same time, you can count the swings to keep track of time.

photosynthesis The process by which plants use the energy of sunlight to make the materials they need to grow.

physicist Scientist who studies the nature of matter and energy.

pineal gland A small gland in the front of the brain that is sensitive to light.

pitch Quality of a sound. Sound waves having a high pitch (like a squeak) have a short wavelength.

planet A large body that orbits a sun. Our solar system has nine planets.

Pluto Ninth planet from the Sun—but sometimes it's closer to the Sun than Neptune is.

pollinate To transfer pollen, a fine powdery substance, from one part of a flower to another in order to fertilize it so that new flowers can grow.

potential energy The energy an object has because of its elevated position.

programmer Someone who writes the instructions that tell computers what to do.

radio telescope A telescope that detects radio waves.

radio wave Light wave with a wavelength too long to be seen.

rain forest Tropical woods with rainfall of at least 100 inches (256 cm) per year.

red giant A star in the later stage of its life when it turns red and becomes very large in size.

red shift The increase in wavelength of light from galaxies that are moving away from us at high speed.

reflect To bounce off.

refraction The change in direction of light when it travels from one medium to another of different density.

relativity A theory by Albert Einstein that describes the peculiar behavior of objects traveling at speeds close to the speed of light.

repel Push apart.

retina A sheet of material in the back of your eye on which the lens of your eye forms an image of things that you see.

revolution The motion of one body about another.

rotation The turning of a body about an axis passing through the body.

satellite A natural or man-made object that orbits a planet. The Moon, which orbits the Earth, is Earth's natural satellite.

Saturn Sixth planet from the Sun.

siphon A tube shaped like an upside-down U that can make water flow from one container to another.

solar Referring to the Sun.

solar eclipse Passage of the Moon between the Sun and the Earth, which casts a shadow on the Earth. Anyone in this shadow would see the Sun blocked out partially or totally.

solar system A star with a group of planets in orbit about it. In addition to the Sun and the nine planets, our solar system consists of moons, asteroids, and comets that orbit it.

solid A form of matter having a fixed shape.

sonar A device submarines use to locate objects underwater from sound waves bounced off the objects.

species A group of organisms that look similar and can breed with each other.

spectrometers Sensitive instruments used to measure small changes in the color or wavelength of light waves.

sphere A round ball.

spiral galaxy A galaxy shaped like our Milky Way.

sundial A device for keeping track of the time based on the position of the Sun's shadow.

supernova A very large explosion of a big star that occurs when it burns most of its fuel.

symmetric Having identical corresponding parts opposite a dividing line or about an axis.

synchronous orbit When a moon or satellite orbits a planet in the same time the planet takes to rotate on its axis, causing the moon to appear stationary in the sky.

telepathy The ability to read other people's minds.

temperature The degree of heat or cold measured using a thermometer; can be based on one of several scales.

tidal force A gravitational force whose strength is different at different places on a body, which tends to deform the body.

tides The movement of the ocean waters that occurs twice each day, caused mainly by the pull of the Moon's gravity.

Titan One of Saturn's moons.

tornado Violent, spinning, funnel-shaped whirlwind.

trillion A thousand billion, written as a one with twelve zeroes after it: 1,000,000,000,000.

Triton One of Neptune's moons.

Turing Test A test developed by Alan Turing, a British computer scientist, to see whether computers can think.

ultrasound Sound waves whose wavelength is too short for humans to hear. Bats use ultrasound echoes to sense their environment.

ultraviolet Light wave with a wavelength too short to be seen.

universe All space and everything in it.

Uranus Seventh planet from the Sun.

vapor Gas form of a liquid.

Venus The second closest planet to the Sun.

volume The size of a three-dimensional object or region of space.

water cycle The continual movement of water from place to place on the Earth.

wavelength The distance between crests of a wave.

weight The force with which an object presses downward due to gravity.

white dwarf A star in the stage of life when it gets very small and hot, and looks whitish in color.

wormholes Black holes that connect to another part of space and time—if they exist.

X-ray Light whose wavelength is much too short to be seen, which can be used to take pictures of the inside of the body.

Index